Cambridge International AS & A Level Mathematics

Probability & Statistics 2

STUDENT'S BOOK

Louise Ackroyd, Yimeng Gu, Sharon McBride

Series Editor: Dr Adam Boddison

William Collins' dream of knowledge for all began with the publication of his first book in 1819.

A self-educated mill worker, he not only enriched millions of lives, but also founded a flourishing publishing house. Today, staying true to this spirit, Collins books are packed with inspiration, innovation and practical expertise. They place you at the centre of a world of possibility and give you exactly what you need to explore it.

Collins. Freedom to teach.

Published by Collins
An imprint of HarperCollins*Publishers*
The News Building
1 London Bridge Street
London
SE1 9GF

HarperCollins*Publishers* Macken House
39/40 Mayor Street Upper
Dublin 1
DO1 C9W8
Ireland

Browse the complete Collins catalogue at
www.collins.co.uk

British Library Cataloguing in Publication Data

A catalogue record for this publication is available from the British Library.

Commissioning editor: Jennifer Hall
In-house editor: Lara McMurray
Authors: Louise Ackroyd/Yimeng Gu/Sharon McBride
Series editor: Dr Adam Boddison
Development editor: Tim Major
Project manager: Emily Hooton
Copyeditor: Gwynneth Drabble
Reviewer: Adele Searle
Proofreader: Joan Miller
Answer checkers: David Hemsley/Emma Thomas
Cover designer: Gordon MacGilp
Cover illustrator: Maria Herbert-Liew
Typesetter: Jouve India Private Ltd
Illustrator: Jouve India Private Ltd/Ken Vail Graphic Design Ltd
Production controller: Sarah Burke
Printed and bound by Ashford Colour Press Ltd

This book contains FSC™ certified paper and other controlled sources to ensure responsible forest management.

For more information visit: www.harpercollins.co.uk/green

Acknowledgements

The publishers wish to thank Cambridge Assessment International Education for permission to reproduce questions from past AS & A Level Mathematics papers. Cambridge Assessment International Education bears no responsibility for the example answers to questions taken from its past papers. These have been written by the authors. Exam-style questions and sample answers have been written by the authors.

The publishers wish to thank the following for permission to reproduce photographs. Every effort has been made to trace copyright holders and to obtain their permission for the use of copyright material. The publishers will gladly receive any information enabling them to rectify any error or omission at the first opportunity.

pvi Cultura Creative (RF)/Alamy Stock Photo, p1 Cultura Creative (RF)/Alamy Stock Photo, p25 bibiphoto/Shutterstock, p46 Andriy Solovyov/ Shutterstock, p70 David R. Frazier Photolibrary, Inc./ Alamy Stock Photo, p100 Hongqi Zhang/Alamy Stock Photo.

Full worked solutions for all exercises, exam-style questions and past paper questions in this book available to teachers by emailing international.schools@harpercollins.co.uk and stating the book title.

CONTENTS

Full worked solutions for all exercises, exam-style questions and past paper questions in this book available to teachers by emailing international.schools@harpercollins.co.uk and stating the book title.

INTRODUCTION

This book is part of a series of nine books designed to cover the content of the Cambridge International AS and A Level Mathematics and Further Mathematics. The chapters within each book have been written to mirror the syllabus, with a focus on exploring how the mathematics is relevant to a range of different careers or to further study. This theme of *Mathematics in life and work* runs throughout the series with regular opportunities to deepen your knowledge through group discussion and exploring real-world contexts.

Within each chapter, examples are used to introduce important concepts. Practice questions are provided to help you to achieve mastery. Developing skills in modelling, problem solving and mathematical communication can significantly strengthen overall mathematical ability. The practice questions in every chapter have been written with this in mind and include symbols to indicate which of these underlying skills are being developed. Exam-style questions are included at the end of each chapter and a bank of real Cambridge past exam questions is included at the end of the book.

A range of other features throughout the series will help to optimise your learning. These include:

> **key information boxes** – highlighting important learning points or key formulae

> **commentary boxes** – tackling potential misconceptions and strengthening understanding through probing questions

> **stop and think** – encouraging independent thinking and developing reflective practice.

Key mathematical terminology is listed at the beginning of each chapter and a glossary is provided at the end of each book. Similarly, a summary of key points and key formulae is provided at the end of each chapter. Where appropriate, alternative solutions are included within the worked solutions to encourage you to consider different approaches to solving problems.

Probability & Statistics 2 builds on the concepts that were introduced in **Probability & Statistics 1**. You will learn how to analyse increasingly complex real-life scenarios, using statistical models and approximations. Much of the content is centred on the use of three particular distributions: the binomial, the normal and the Poisson. However, this book will also introduce you to distributions that are defined by mathematical equations, providing real-life contexts in which your knowledge of integration can be applied.

Testing whether or not a particular intervention has had a significant impact is a useful skill in a range of careers, such as conducting randomised controlled trials for new medicines or evaluating the impact of new teaching methods on students' examination results. Hypothesis testing is introduced in this book and you will learn how it can be used to evaluate claims of change or improvement.

FEATURES TO HELP YOU LEARN

Mathematics in life and work

Each chapter starts with real-life applications of the mathematics you are learning in the chapter to a range of careers. This theme is picked up in group discussion activities throughout the chapter.

Learning objectives

A summary of the concepts, ideas and techniques that you will meet in the chapter.

Language of mathematics

Discover the key mathematical terminology you will meet in this chapter. As you work through the chapter, key words are written in bold. The words are defined in the glossary at the back of the book.

Prerequisite knowledge

See what mathematics you should know before you start the chapter, with some practice questions to check your understanding.

Explanations and examples

Each section begins with an explanation and one or more worked examples, with commentary, where appropriate, to help you follow. Some show alternative solutions in the accompanying commentary to get you thinking about different approaches to a problem.

---

1 THE POISSON DISTRIBUTION

Mathematics in life and work

In this chapter you will learn about the Poisson distribution and how to use it to calculate probabilities of the number of events happening in a set period of time. The modelling of situations as a Poisson distribution can be used in many different careers, for example:

› If you were a doctor and you knew how widespread a particular disease was, you could work out the likelihood of one or more patients having the disease. This would allow you to make decisions about the medication required for your patients.

› If you worked within a quality control department in a factory, you could use the Poisson distribution to model the number of times a machine breaks down in a month. You could then work out the probability of there being more than two breakdowns in a month and put measures in place to minimise the likelihood.

› If you were a road traffic controller, you might use historical data to calculate the probability of there being three or more accidents on a motorway on a single day.

This chapter focuses on how scientists can use the Poisson distribution.

LEARNING OBJECTIVES

You will learn how to:

› calculate probabilities for the Poisson distribution
› use the fact that if $X \sim Po(\lambda)$ then the mean and variance of X is λ
› use the Poisson distribution as a model
› use the Poisson distribution to approximate the binomial distribution
› use the normal distribution to approximate the Poisson distribution.

LANGUAGE OF MATHEMATICS

Key words and phrases you will meet in this chapter:

› binomial distribution, continuity correction, normal distribution, Poisson distribution

---

Example 4

The data about the amount of mail a person received was recorded in the table below. Find the mean and variance, and hence comment on the suitability of a Poisson distribution model.

Number of letters per day	0	1	2	3	4
Frequency	27	29	16	6	2

Solution

$$\mu = \frac{\sum xf}{\sum f}$$

Colour-coded questions

Questions are colour-coded (green, blue and red) to show you how difficult they are. Exercises start with more accessible (green) questions and then progress through intermediate (blue) questions to more challenging (red) questions.

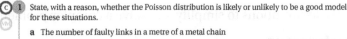

1 State, with a reason, whether the Poisson distribution is likely or unlikely to be a good model for these situations.

   a The number of faulty links in a metre of a metal chain

   b The number of cars passing a school in 10-minute intervals from midnight to 11 am

   c The number of errors in a randomly chosen page of a book

   d The number of giraffes spotted on a 3-hour safari

   e The number of dandelions (a type of wild flower) in a randomly chosen square metre of field.

2 Given the random variable $X \sim \text{Po}(2.5)$, find:

   a $P(X = 4)$ **b** $P(X = 2)$ **c** $P(X \leqslant 3)$.

3 The random variable Y has a Poisson distribution with a mean of 7.2. Find:

   a $P(Y = 6)$ **b** $P(5 < Y \leqslant 9)$ **c** $P(Y \geqslant 4)$.

Question-type indicators

The key concepts of problem solving, communication and mathematical modelling underpin your A level Mathematics course. You will meet them in your learning throughout this book and they underpin the exercises and exam-style questions. All mathematics questions will include one or more of the key concepts in different combinations. We have labelled selected questions that are especially suited to developing one or more of these key skills with these icons:

(PS) **Problem solving** – mathematics is fundamentally problem solving and representing systems and models in different ways. These include: algebra, geometrical techniques, calculus, mechanical models and statistical methods. This icon indicates questions designed to develop your problem-solving skills. You will need to think carefully about what knowledge, skills and techniques you need to apply to the problem to solve it efficiently.

These questions may require you to:

› use a multi-step strategy

› choose the most efficient method, or

› bring in mathematics from elsewhere in the curriculum

› look for anomalies in solutions

› generalise solutions to problems.

(C) **Communication** – communication of steps in mathematical proof and problem solving needs to be clear and structured, using algebra and mathematical notation, so that others can follow your line of reasoning. This icon indicates questions designed to develop your mathematical communication skills. You will need to structure your solution clearly, to show your reasoning and you may be asked to justify your conclusions.

These questions may require you to:

› use mathematics to demonstrate a line of argument

› use mathematical notation in your solution

› follow mathematical conventions to present your solution clearly

› justify why you have reached a conclusion.

(MM) **Mathematical modelling** – a variety of mathematical content areas and techniques may be needed to turn a real-world situation into something that can be interpreted through mathematics. This icon indicates questions designed to develop your mathematical modelling skills. You will need to think carefully about what assumptions you need to make to model the problem, and how you can interpret the results to give predictions and information about the real world.

These questions may require you to:

› construct a mathematical model of a real-life situation, using a variety of techniques and mathematical concepts

> use your model to make predictions about the behaviour of mathematical systems

> make assumptions to simplify and solve a complex problem.

Key information

These boxes highlight information that you
need to pay attention to and learn, such as
key formulae and learning points

Stop and think

Stop and think   What other method could you use?

These boxes present you
with probing questions and problems to help you to reflect on what you have been learning.
They challenge you to think more widely and deeply about the mathematical concepts, tackle
misconceptions and, in some cases, generalise beyond the syllabus. They can be a starting point
for class discussions or independent research. You will need to think carefully about the question
and come up with your own solution.

Mathematics in life and work – Group discussions give you the chance to apply the
skills you have learned to a model of a real-life maths problem, from a career that uses
maths. Your focus is on applying and practising the concepts, and coming up with your own
solutions, as you would in the workplace. These tasks can be used for class discussions, group
work or as an independent challenge.

Summary of key points

At the end of each chapter, there is a summary of key formulae and learning points.

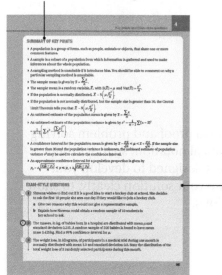

Exam-style questions

Practise what you have learnt throughout the chapter
with questions written in examination style by our
authors, progressing in order of difficulty.

The last **Mathematics in life and work** question draws
together the skills that you have gained in this chapter
and applies them to a simplified real-life scenario.

At the end of the book, test your mastery of what you have learned in the **Summary review**
section. Practise the basic skills with some Probability & Statistics 1 past paper questions, and
then go on to try carefully selected questions from Cambridge International A Level past exam
papers and exam-style questions on new topics. Extension questions, written by our authors,
give you the opportunity to challenge yourself and prepare you for more advanced study.

1 THE POISSON DISTRIBUTION

Mathematics in life and work

In this chapter you will learn about the Poisson distribution and how to use it to calculate probabilities of the number of events happening in a set period of time. The modelling of situations as a Poisson distribution can be used in many different careers, for example:

> If you were a doctor and you knew how widespread a particular disease was, you could work out the likelihood of one or more patients having the disease. This would allow you to make decisions about the medication required for your patients.

> If you worked within a quality control department in a factory, you could use the Poisson distribution to model the number of times a machine breaks down in a month. You could then work out the probability of there being more than two breakdowns in a month and put measures in place to minimise the likelihood.

> If you were a road traffic controller, you might use historical data to calculate the probability of there being three or more accidents on a motorway on a single day.

This chapter focuses on how scientists can use the Poisson distribution.

LEARNING OBJECTIVES

You will learn how to:

> calculate probabilities for the Poisson distribution

> use the fact that if $X \sim \text{Po}(\lambda)$ then the mean and variance of X is λ

> use the Poisson distribution as a model

> use the Poisson distribution to approximate the binomial distribution

> use the normal distribution to approximate the Poisson distribution.

LANGUAGE OF MATHEMATICS

Key words and phrases you will meet in this chapter:

> binomial distribution, continuity correction, normal distribution, Poisson distribution

PREREQUISITE KNOWLEDGE

You should already know how to:

> calculate exponential functions and factorials

> calculate the mean and variance from a frequency table

> calculate the mean and variance of a binomial distribution

> apply a continuity correction to allow the use of a normal distribution to approximate a discrete random variable.

You should be able to complete the following questions correctly:

1 Calculate the following.

   a e^3 **b** e^{-2} **c** $\dfrac{5!}{3!}$ **d** $4! \times e^{-0.5}$

2 Find the mean and variance.

X	21	23	26	27
Frequency	2	5	1	3

3 Given the random variable $X \sim B(10, 0.3)$, find:

   a $E(X)$ **b** $Var(X)$.

4 X is a discrete random variable that can be approximated by a continuous random variable Y. Use a continuity correction to write these probabilities in terms of Y.

   a $P(X < 35)$ **b** $P(X \geqslant 27)$ **c** $P(23 \leqslant X < 44)$ **d** $P(X = 4)$

1.1 The Poisson distribution

A company keeps a record of how many phone calls are received by their customer service department. They record the number of phone calls in 10-minute blocks. The diagram on the right shows when phone calls were received during the first hour in each 10-minute interval.

2   1   3   0   2   1

The table below shows the number of calls in each 10-minute block over a day.

Number of calls in 10 minutes	Frequency
0	6
1	12
2	13
3	9
4	5
5	2
6	1
⩾7	0

The mean number of phone calls in 10 minutes is

$$\frac{\sum xf}{\sum f} = \frac{0 \times 6 + 1 \times 12 + 2 \times 13 + 3 \times 9 + 4 \times 5 + 5 \times 2 + 6 \times 1}{6 + 12 + 13 + 9 + 5 + 2 + 1}$$

$$= 2.10 \text{ (3 s.f.)}$$

The random variable R can be used to represent the number of phone calls made to the customer service line in 10 minutes. The probability distribution of R is given in the table below.

r	0	1	2	3	4	5	6	$\geqslant 7$
$P(R = r)$	0.125	0.25	0.271	0.188	0.104	0.417	0.0208	0

The probability of receiving zero calls in 10 minutes is:

$$P(R = 0) = \frac{6}{48} = 0.125$$

This is a real-life example of a Poisson distribution, named after the French mathematician Siméon Denis Poisson (1781–1840). The Poisson distribution can model a discrete random variable and is concerned with the number of times a random event happens in a given interval. You can use a Poisson distribution when the events:

A random event happens in a given interval means that no two events can happen at precisely the same point in time or space.

> occur independently

> can only happen singly

> occur randomly

> occur at a constant mean rate.

In the example above, the random variable R can be modelled by a Poisson distribution with mean 2.1. This can be written as $R \sim Po(2.1)$.

The probability of r phone calls in 10 minutes can be calculated using the formula $P(R = r) = \dfrac{e^{-\lambda}\lambda^r}{r!}$

KEY INFORMATION

A Poisson distribution can be written as $X \sim Po(\lambda)$ where λ (lambda) is the constant mean rate.

The probability that exactly r events will occur in a particular interval is:

$$P(R = r) = \frac{e^{-\lambda}\lambda^r}{r!}$$

for $r = 0, 1, 2, 3, \ldots$

The theoretical probabilities for the discrete random variable R are shown below and can be compared with the table above.

r	0	1	2	3	4	5	6
$P(R = r)$	$e^{-2.1}\dfrac{2.1^0}{0!}$ $= 0.1225$	$e^{-2.1}\dfrac{2.1^1}{1!}$ $= 0.2572$	$e^{-2.1}\dfrac{2.1^2}{2!}$ $= 0.2700$	$e^{-2.1}\dfrac{2.1^3}{3!}$ $= 0.1890$	$e^{-2.1}\dfrac{2.1^4}{4!}$ $= 0.0992$	$e^{-2.1}\dfrac{2.1^5}{5!}$ $= 0.0417$	$e^{-2.1}\dfrac{2.1^6}{6!}$ $= 0.0146$

Under the Poisson distribution model, the discrete random variable can be taken by any non-negative integer values.

$0! = 1$

The probability density graph of a Poisson distribution X ~ Po(λ)

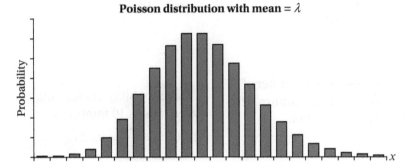

Poisson distribution with mean = λ

As the value of X gets further away from the mean, the probability tends to zero.

> Under the binomial distribution X~B (n, p), the random variable can only be the positive integer that is less than n.

Example 1

A scientist is looking at the number of limpets (a type of shellfish that cling to rocks) on a beach. He finds that there are on average 3 limpets in every 25 cm^2 area.

a   What assumptions have to be made to model this as a Poisson distribution?

b   Find the probability of there being exactly 5 limpets in a randomly chosen 25 cm^2 area.

c   Find the probability there are at least 2 limpets in a randomly chosen 25 cm^2 area.

Solution

a   The limpets must be independently arranged on the rock.

The average number of limpets must be constant.

The limpets must not overlap each other (they must appear singly).

b X ~ Po(3)

$$P(X = 5) = P_5 = \frac{e^{-3} \times 3^5}{5!}$$

$$= 0.100818\ldots$$

$$= 0.101 \text{ (3 s.f.)}$$

c   Probability that at least two is P($X \geqslant 2$)
   $= P(X = 2) + P(X = 3) + P(X = 4) + \ldots$

Since the random variable X has no upper limit, you need to use the fact that all the probabilities sum to 1.

$$P(X \geqslant 2) = 1 - P(X < 2) = 1 - (P(X = 0) + P(X = 1))$$

$$= 1 - \left(\frac{e^{-3} \times 3^0}{0!} + \frac{e^{-3} \times 3^1}{1!} \right)$$

$$= 1 - (0.04978\ldots + 0.14936\ldots)$$

$$= 0.801 \text{ (3 s.f.)}$$

> You can also write P($X \geqslant 2$) = $1 - P(X \leqslant 1)$, as the random variable is discrete.

Example 2

A machine makes fabric. It is known that the mean number of faults in 1 metre is 0.3. Find the probability there are:

a 2 faults in a metre   **b** 4 faults in 5 metres.

Solution

a X = number of faults in 1 metre so $X \sim \text{Po}(0.3)$.

$$P_2 = e^{-0.3}\frac{0.3^2}{2!}$$

$$= 0.0333 \text{ (3 s.f.)}$$

b Y = the number of faults in 5 metres.

Since the mean number of faults is consistent then you know the mean number of faults in 5 metres will be $5 \times 0.3 = 1.5$ so $Y \sim \text{Po}(1.5)$.

$$P_4 = e^{-1.5}\frac{1.5^4}{4!}$$

$$= 0.0471 \text{ (3 s.f.)}$$

> **KEY INFORMATION**
>
> If random variables X_1, X_2, X_3, ... follow Poisson distributions with mean λ_1, λ_2, λ_3, ... respectively, then the random variable, $Y = X_1 + X_2 + X_3 + ...$ also follows a Poisson distribution with mean $\lambda = \lambda_1 + \lambda_2 + \lambda_3 + ...$

Example 3

A printing fault occurs on a fabric at a rate of 0.5 faults per square metre. Emma buys 10 pieces of such fabric, each measuring 5 square metres. What is the probability that six randomly chosen pieces of fabric contain at least three printing faults?

Solution

First, find the probability of a piece of fabric of area 5 square metres that contains more than three printing faults.

X = the number of printing faults, $\lambda = 0.5 \times 5 = 2.5$

$X \sim \text{Po}(2.5)$

$P(X \geq 3) = 1 - P(X \leq 2)$

$$= 1 - P(X = 0) - P(X = 1) - P(X = 2)$$

$$= 1 - e^{-2.5}\frac{2.5^0}{0!} - e^{-2.5}\frac{2.5^1}{1!} - e^{-2.5}\frac{2.5^2}{2!}$$

$$= 1 - 0.082\,08\ldots - 0.205\,21\ldots - 0.256\,51\ldots$$

$$= 0.456 \text{ (3 s.f.)}$$

Then consider a binomial distribution.

Y = the piece of fabric contains more than 3 printing faults

$Y \sim \text{B}(10, 0.4562)$

$P(Y = 6) = {}^{10}_{6}C\, 0.4562^6(1 - 0.4562)^4 = 0.166 \text{ (3 s.f.)}$

Exercise 1.1A

 1 State, with a reason, whether the Poisson distribution is likely or unlikely to be a good model for these situations.

    a The number of faulty links in a metre of a metal chain

    b The number of cars passing a school in 10-minute intervals from midnight to 11 am

    c The number of errors in a randomly chosen page of a book

    d The number of giraffes spotted on a 3-hour safari

    e The number of dandelions (a type of wild flower) in a randomly chosen square metre of field.

2 Given the random variable $X \sim \mathrm{Po}(2.5)$, find:

    a $\mathrm{P}(X = 4)$ **b** $\mathrm{P}(X = 2)$ **c** $\mathrm{P}(X \leqslant 3)$.

3 The random variable Y has a Poisson distribution with a mean of 7.2. Find:

    a $\mathrm{P}(Y = 6)$ **b** $\mathrm{P}(5 < Y \leqslant 9)$ **c** $\mathrm{P}(Y \geqslant 4)$.

4 The number of defects in a pane of glass has a mean of 0.6. Find the probability of:

    a 1 defect   **b** no defects   **c** more than 2 defects.

5 The number of customers using an ATM in 5 minutes can be modelled by a Poisson distribution with mean of 1.1. Find the probability of:

    a 3 customers arriving in 5 minutes

    b more than 2 customers arriving in 5 minutes

    c fewer than 5 customers using the ATM in a 30-minute period.

 6 A shop stocks a local newspaper. They buy 6 copies of it a week. On average, in a randomly chosen week, they sell 4 copies.

    a State the assumptions needed to use a Poisson distribution.

    b Find the probability that more customers want to buy the newspaper than they have copies in stock.

7 On a stretch of road there are on average 1.8 breakdowns per day. Stating any assumptions, find:

    a the probability there are fewer than 3 breakdowns on a randomly picked day

    b the probability there are more than 5 breakdowns in a week

    c the probability that there are more than 5 breakdowns each week in a month.

8 The sales of children's bikes at two shops from the same chain follow an independent Poisson distribution with mean 5 per day at the first shop and 3 per day at the second shop. Find the probability that, on a random day:

    a the first shop sells fewer than 3 bikes

    b the second shop sells fewer than 3 bikes

    c the total sales by both shops is from 4 to 6 bikes.

  C Communication   **MM** Mathematical modelling   **PS** Problem solving

9 A photocopier jams on average 1.5 times over 2 hours.

  a Find the probability that the photocopier jams:

    i at least four times over a 5-hour period

    ii exactly 5 times in a working day of 8 hours.

  b Over a period of 5 working days what is the probability that the photocopier jams exactly five times on each of two different days?

10 People go to the cinema either alone or as part of a group.

On a particular night, the number of people who arrive alone at the cinema during a 1-minute interval, X, can be modelled by a Poisson distribution with mean 4.

  a Find the probability of exactly 3 people arriving alone during a randomly chosen minute.

The number of groups who arrive at the cinema during a 1-minute interval, Y, may also be modelled by a Poisson distribution with mean 2.5.

  b Find the probability that exactly 3 groups arrive within a randomly chosen 2-minute interval.

  c The ticket office is open from 9:00 am to 12:00 noon. What is the probability that the event, exactly 3 groups arrive in 2 minutes, happens exactly 10 times in a day?

> **Stop and think**
>
> Under certain circumstances, the number of customers to a shop can be modelled as a Poisson distribution. When might a Poisson distribution be a good model? When would it be a bad model?

The mean and variance of a Poisson distribution

Since λ is the average rate of occurrence of an event in a set time or space, it is also the mean and the expectation of the Poisson distribution. The variance of a Poisson distribution is also equal to λ. This is a special property of the Poisson distribution.

KEY INFORMATION

If $X \sim \text{Po}(\lambda)$, then $\mu = \lambda$ and $\sigma^2 = \lambda$.

From **Mathematics: Probability & Statistics 1, Chapter 3, Discrete Random Variables**, $\text{Var}(X) = \text{E}(X^2) - (\text{E}(X))^2$.

Since $\text{E}(X) = \lambda$:

$\text{Var}(X) = \text{E}(X^2) - (\text{E}(X))^2$

$\qquad = \text{E}(X(X-1) + X) - (\text{E}(X))^2$ Rewrite X^2 in terms of X.

$\qquad = \text{E}(X(X-1)) + \text{E}(X) - (\text{E}(X))^2$

$\qquad = \text{E}(X(X-1)) + \lambda - \lambda^2$

$\qquad = \sum_{x=0}^{\infty} (x)(x-1)\frac{e^{-\lambda}\lambda^x}{x!} + \lambda - \lambda^2$

$\qquad = \sum_{x=2}^{\infty} (x)(x-1)\frac{e^{-\lambda}\lambda^x}{x!} + \lambda - \lambda^2$ The first two terms are zero when $x = 0$ and $x = 1$.

$$= \sum_{x=2}^{\infty} \frac{e^{-\lambda}\lambda^x}{(x-2)!} + \lambda - \lambda^2$$

Simplify

$$(x)(x-1)\frac{e^{-\lambda}\lambda^x}{x(x-1)(x-2)!}$$

$$= e^{-\lambda}\lambda^2 \sum_{x=2}^{\infty} \frac{\lambda^{x-2}}{(x-2)!} + \lambda - \lambda^2$$

$$= e^{-\lambda}\lambda^2 \left(\frac{\lambda^0}{0!} + \frac{\lambda^1}{1!} + \frac{\lambda^2}{2!} + \ldots \right) + \lambda - \lambda^2$$

Factor out $e^{-\lambda}\lambda^{x-2}$

$$= e^{-\lambda}\lambda^2 e^{\lambda} + \lambda - \lambda^2$$

$$= \lambda^2 + \lambda - \lambda^2$$

$$= \lambda$$

This shows that if the mean of a Poisson distribution is λ, then the variance is also λ, and the standard deviation is $\sqrt{\lambda}$.

Example 4

The data about the amount of mail a person received was recorded in the table below. Find the mean and variance, and hence comment on the suitability of a Poisson distribution model.

Number of letters per day	0	1	2	3	4
Frequency	27	29	16	6	2

Solution

$$\mu = \frac{\sum xf}{\sum f}$$

$$= \frac{0 \times 27 + 1 \times 29 + 2 \times 16 + 3 \times 6 + 4 \times 2}{27 + 29 + 16 + 6 + 2}$$

$$= 1.0875$$

In this case, x is the number of letters per day and f is the frequency.

$$\sigma^2 = \frac{\sum x^2 f}{\sum f} - \mu^2$$

$$= \frac{0^2 \times 27 + 1^2 \times 29 + 2^2 \times 16 + 3^2 \times 6 + 4^2 \times 2}{27 + 29 + 16 + 6 + 2} - 1.0875^2$$

$$= 1.055$$

Since $1.0875 \approx 1.055$ (when rounded to 1 d.p.), we can say the Poisson distribution is a good model.

Example 5

The mean of a discrete random variable, X, is 3.7 and it can be modelled by a Poisson distribution.

a Write down the variance and standard deviation of X.

b Find the probability that X is within 1 standard deviation of the mean.

Solution

a Since the mean and variance of a Poisson distribution
$$\mu = \sigma^2 = 3.7$$

$$\sigma = \sqrt{3.7} = 1.92 \ (3 \text{ s.f.})$$

b The probability that X is within 1 standard deviation of the mean can be written as the mean ± standard deviation, so:

$$P(\mu - \sigma < X < \mu + \sigma)$$

$$= P(3.7 - 1.92 < X < 3.7 + 1.92)$$

$$= P(1.78 < X < 5.62)$$

$$= P(X = 2, 3, 4, 5)$$

$$= \frac{e^{-3.7} \times 3.7^2}{2!} + \frac{e^{-3.7} \times 3.7^3}{3!} + \frac{e^{-3.7} \times 3.7^4}{4!} + \frac{e^{-3.7} \times 3.7^5}{5!}$$

$$= e^{-3.7}\left(\frac{3.7^2}{2!} + \frac{3.7^3}{3!} + \frac{3.7^4}{4!} + \frac{3.7^5}{5!} \right)$$

$$= 0.713\,88\ldots$$

$$= 0.714 \ (3 \text{ s.f.})$$

> Remember that X is discrete in the Poisson distribution and it only takes non-negative integer values.

> Factoring out $e^{-3.7}$ will reduce the possibility of keying errors. It will also make use of the calculator more efficient.

Exercise 1.1B

1 For each of the following data sets:

  i find the mean and variance

  ii decide whether a Poisson distribution is a good model.

a

X	0	1	2	3	4	5	6
Frequency	6	13	14	9	5	2	1

b

Y	0	1	2	3	4	5	6	7
Frequency	5	12	16	20	14	3	2	1

c

N	0	1	2
Frequency	32	13	3

d

M	19	20	21	22	23
Frequency	2	8	12	5	1

e

V	0	1	2	3	4	5	6	7
Frequency	5	11	17	17	13	8	3	2

2 A discrete random variable X has a Poisson distribution with mean 4.

   a Write down the variance and standard deviation.

   b Calculate $P(X < \mu)$.

3 The random variable $Y \sim Po(5.1)$.

   a Write down the standard deviation.

   b Find $P(X < \mu - \sigma)$.

4 The expected value of a discrete random variable, X, is 5.2 and it can be modelled by a Poisson distribution.

   a State the mean and variance of X.

   b Find the probability that X is within 1 standard deviation of the mean.

5 The numbers of holiday packages sold by an online travel agency in twelve successive hours on a particular day were:

   15   9   16   11   23   18   27   14   29   22   5   16

   a Calculate the mean and the variance of the data.

   b State, with a reason based on the calculations, whether it is likely that the Poisson distribution will provide a suitable model for the data.

PS **6** The number of accidents in a factory can be modelled as a Poisson distribution. From historical data it is known that there are on average 3.2 accidents per week.

   a State the value of λ and σ.

   b Find the probability the number of accidents in a week is within 1.5 standard deviations of the mean.

7 The sales of tennis racquets in a sports shop may be modelled by a Poisson distribution with a mean of 1.6 sales per day.

   a State the mean and variance of tennis racquets sold in 7 days.

   b Find the probability that the number of tennis racquets sold in 7 days is at least 2 standard deviations less than the mean.

8 A call centre receives an average of 3.7 calls per minute between 8 am and 5 pm.

  a Find the probability that there will be more than 4 calls in a randomly chosen minute.

  b The manager has a team of 5 staff. Suggest, with a reason, whether or not she should increase the number of staff on the team to 6.

9 A summer holiday camp is advertised 4 weeks before it is due to take place. Throughout these 4 weeks, the number of places booked follow a Poisson distribution with mean 3 per day.

  a Find the probability that, during the first week, exactly 20 places are booked.

  b Find the probability that the number of places booked in a particular day exceeds the mean by one standard deviation.

The organiser then decided that in the first 2 weeks, if the places booked exceeds the mean by one standard deviation on more than 4 days, they would put on a live entertainment.

  c Find the probability that the organiser will put on a live entertainment.

10 In a basketball match, a player can score one, two or three points depending on shooting location. During a particular match, the number of three-point scores may be modelled by a Poisson distribution with mean 1.5 per 5 minutes.

  a Find the probability that the number of three-point scores exceeds the mean by 2 standard deviations in a randomly chosen minute.

In another match, Mr Ali recorded the score during a 10-minute interval. It was as follows:

2 1 3 2 1 3 2 1 1 1 2 1 3 1 2 1 2 2 3 2

  b Find the average number of two-point scores per minute over a 10-minute interval.

  c Find the probability that two-points are scored ten times during a 15-minute period.

Mathematics in life and work: Group discussion

You are monitoring an active volcano that is on the edge of a town. The town's authorities are developing emergency plans in case there is a big eruption and have asked for your help in modelling the number of eruptions to help inform their plans.

1 How could you work out the average rate of eruptions in a year?

From historical data you learn that there are on average 2 eruptions every 5 years.

2 Find the probability of there being more than 4 eruptions in 10 years.

3 How could the people of the town use your findings to help plan for volcano eruptions?

1.2 Using the Poisson distribution to approximate the binomial distribution

In **Probability & Statistics 1, Chapter 4, The normal distribution** you learnt that a normal distribution can be used to approximate a binomial distribution when $np \geqslant 5$ and $nq \geqslant 5$.

The Poisson distribution can be used to approximate a binomial distribution if the mean and variance are approximately the same.

If $X \sim B(n, p)$, then

$$E(X) = \mu = np$$

$$Var(X) = \sigma^2 = np(1 - p)$$

So, if n is large and p is small, $np(1 - p)$ is approximate to np. This implies that the mean and variance are close enough to each other to allow the binomial distribution to be approximated by a Poisson distribution with the same mean ($\mu = \lambda = np$).

> As a guide, n is considered to be large when $n > 50$ and p is considered to be small if $np < 5$.

> **KEY INFORMATION**
>
> When $n > 50$ and $np < 5$ then the binomial distribution $B(n, p)$ can be approximated by a Poisson distribution $Po(np)$.

Example 6

1.5% of the population of a town have a particular type of allergy. A random sample of 120 people are asked whether they have this allergy. By considering a suitable approximation, find the probability that:

a exactly 3 people have the allergy

b at least 3 people have the allergy.

Solution

a $X \sim B(120, 0.015)$

$n = 120$ which is greater than 50 and $np = 120 \times 0.015 = 1.8$ which is less than 5, therefore X can be approximated by $Po(1.8)$.

$$P(X = 3) = e^{-1.8} \frac{1.8^3}{3!}$$

$$= 0.160\,67\ldots$$

$$= 0.161 \ (3 \text{ s.f.})$$

b $P(X \geqslant 3) = 1 - P(X \leqslant 2)$

$$= 1 - (P(X = 0) + P(X = 1) + P(X = 2))$$

$$= 1 - e^{-1.8}\left(\frac{1.8^0}{0!} + \frac{1.8^1}{1!} + \frac{1.8^2}{2!}\right)$$

$$= 0.269\,37\ldots$$

$$= 0.269 \ (3 \text{ s.f.})$$

Example 7

A gardener has found that a particular brand of poppy seeds germinate 97% of the time. He plants 100 randomly selected seeds in the garden. Using a suitable approximation, find the probability that exactly 95 seeds germinate.

Solution

Let X be the number of seeds that germinate:

$X \sim B(100, 0.97)$

$n = 100$ so is greater than 50 but $np = 100 \times 0.97 = 97$ which is too large for a Poisson approximation.

Since p is closer to 1, you can redefine the random variable as the complementary variable. This makes p small.

Let Y be the number of seeds that do not germinate:

$Y \sim B(100, 0.03)$

p is now closer to 0, ($np = 100 \times 0.03 = 3$) so a Poisson approximation is appropriate with mean 3, Po(3).

$P(X = 95) = P(Y = 5)$

$$= e^{-3}\frac{3^5}{5!}$$

$$= 0.101 \text{ (3 s.f.)}$$

> The complementary variable considers the opposite of the variable. In this case, the complementary variable is the number of seeds that do not germinate. You can calculate the probability of this as $1 - p$.

Stop and think   In what circumstances might it be more appropriate to use a Poisson approximation to a binomial distribution, than to use the binomial distribution directly?

Exercise 1.2A

For each of these questions, use a suitable approximation where appropriate.

1 The random variable $X \sim B(150, 0.02)$.

   a Find the mean of X.

   b Using a suitable approximation, find $P(X = 2)$.

2 X is a random variable with distribution B(60, 0.07). Using a suitable approximation, find:

   a $P(X < 2)$ **b** $P(X \geqslant 4)$

3 A factory manufactures pens in batches of 500. It is known that the probability of a randomly selected pen being faulty is 0.007.

  a State a suitable approximation and justify your answer.

  b Find the probability that in a randomly chosen batch:

    i 3 pens are faulty   **ii** at most, 4 pens are faulty.

4 When two fair cubical dice are thrown, a 'double 1' is when both dice show a 1.

  a Find the probability of getting a double 1.

The two dice are rolled together 140 times.

  b What is the probability that they landed on a double 1 at least 4 times?

5 At a funfair, 100 children made five throws with a fair coin. Using a suitable approximation, calculate the probability that four children obtain five heads.

6 Charity shops receive donations from local residents. On average, one out of 25 donations contains soft toys. A charity shop received 60 donations over a period of time. Using a suitable approximation, find the probability that exactly six donations contain soft toys.

7 In a doctor's surgery, 1.2% of patients miss their appointments. A doctor has 145 appointments available in a week. In a particular week 80% of appointment slots are booked. Find:

  a the expected number of people who miss their appointment in this week

  b the probability that at least 5 people miss their appointments in the week.

8 In a school it is known that 91.3% of people are right-handed. In one year group there are 54 students. Let X be the number of right-handed students in the year group.

  a Comment on the suitability of a Poisson approximation of X.

  b Using a suitable approximation, find the probability that there are more than 48 right-handed students in the year group.

9 Sophie mixed one whole peanut into cookie dough to make twelve cookies. A random cookie is chosen. Using a suitable approximation, find the probability that the chosen cookie contains the peanut.

10 Amy and Beth play each other at table tennis. For each game, the probability that Amy wins is 0.8.

  a Find the probability that in 12 games, Amy wins exactly 4 games.

Amy then attended a training course. After completing the training course, the probability that Amy wins each game increased to 0.92.

Amy and Beth agree to play a further fifty games.

  b Using a suitable approximation, find the probability that Amy wins exactly 48 games.

1.3 Using the normal distribution to approximate the Poisson distribution

The diagrams below illustrate that as λ increases, the distribution becomes less skewed and closer to being symmetrical.

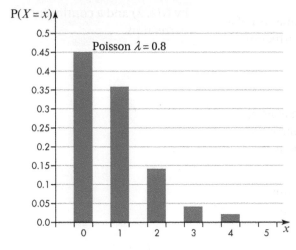

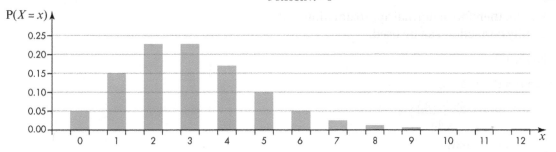

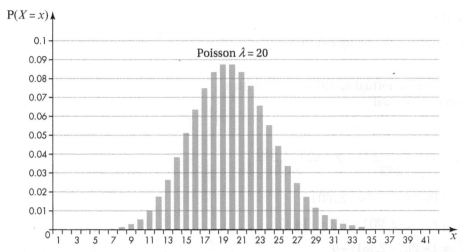

When $\lambda > 15$ the Poisson distribution has the bell shape of a normal distribution with mean λ and variance λ. This means you can approximate the Poisson distribution by a normal distribution $N(\lambda, \lambda)$.

By convention, λ is large enough to produce a symmetrical distribution when $\lambda > 15$.

In **Probability & Statistics 1, Chapter 4, The normal distribution** you saw that when a normal is used to approximate the binomial distribution you must use a continuity correction. This is because you are using a continuous distribution as an approximation to a discrete distribution.

So, when you use a normal distribution as an approximation of a Poisson distribution you must also use a continuity correction.

> **KEY INFORMATION**
>
> If $X \sim \text{Po}(\lambda)$ where $\lambda > 15$, then X can be approximated by $\text{N}(\lambda, \lambda)$ and a continuity correction.

Example 8

A physicist is measuring the radioactive decay of an isotope. It is known that the isotope decays by emitting 45 particles a second on average, and that it can be modelled using a Poisson distribution. Find the probability that:

a the isotope emits fewer than 50 particles in a second

b the isotope emits between 150 and 400 particles in 10 seconds.

Solution

a $X \sim \text{Po}(45)$, therefore a normal approximation with a continuity correction can be used.

$Y \sim \text{N}(45, 45)$

$$P(X < 50) = P(Y \leqslant 49.5)$$

$$= P\left(Z < \frac{49.5 - 45}{\sqrt{45}}\right)$$

$$= P(Z < 0.671)$$

$$= 0.749 \text{ (3 s.f.)}$$

b In 10 seconds the rate of particles being emitted is
$45 \times 10 = 450$

Since this is greater than 15 a normal approximation with a continuity correction can be used.

$W \sim \text{N}(450, 450)$

$$P(150.5 < W < 399.5) = P\left(\frac{150.5 - 450}{\sqrt{450}} < Z < \frac{399.5 - 450}{\sqrt{450}}\right)$$

$$= P(-14.12 < Z < -2.381)$$

$$= P(Z < -2.381) - P(Z < -14.12)$$

$$= 1 - P(Z < 2.381) - 0$$

$$= 1 - 0.9913$$

$$= 0.0087$$

> **Stop and think**   Why might a normal approximation to the Poisson distribution be particularly useful when considering a probability range, such as that in **Example 8 part b**?

Exercise 1.3A

For each of these questions, use a suitable approximation where appropriate.

1 Given the random variable $X \sim \text{Po}(19)$, find:

   a $\text{P}(X \leqslant 20)$ **b** $\text{P}(X > 13)$

   c $\text{P}(10 \leqslant X < 25)$ **d** $\text{P}(X = 19)$.

2 On a stretch of road there are 20 accidents per year on average. Find the probability that in a randomly chosen year there are:

   a fewer than 10 accidents   **b** at least 23 accidents

   c more than 16 but less than 25 accidents

MM 3 A football teams scores on average 1.2 goals per match. Find the probability that, in a season of 28 matches, the team scored more than 40 goals.

4 An employee in a clothes factory inserts zips into trousers. It is known that the number of zips they can insert in 1 hour follows a Poisson distribution with mean 10.3.

   a Find the probability that the employee inserts 11 zips in an hour.

   The factory employees work in shifts. Each shift last for 8 hours.

   b Calculate the probability that the employee inserts more than 100 zips during a shift.

5 In a random sample of water from a lake it is found that there are 11 bacteria per ml on average. Assuming the number of bacteria can be modelled by a Poisson distribution, calculate the probability that 10 ml of pond water contains:

   a from 85 to 115 bacteria (inclusive)   **b** more than 90 bacteria   **c** exactly 110 bacteria.

6 Given that $X \sim \text{Po}(40)$ and $\text{P}(X < x) \leqslant 0.1$, find the largest possible integer value of x.

7 **a** The random variable $A \sim \text{Po}(36)$. Use a suitable approximation to find $\text{P}(A > 40)$.

   b An experiment consists of 500 trials. For each trial, the probability that the result is a success is 0.962, independent of all other trials. The total number of successes is denoted by S.

     i Explain why the distribution of S cannot be well approximated by a Poisson distribution.

     ii Use an appropriate approximation to find the probability that the total number of successes is greater than 490.

8 The number of arrivals per minute at a restaurant is modelled by a Poisson distribution. During a Friday evening, $\lambda = 0.73$.

   a Give one reason why the proposed Poisson distribution might be a suitable model.

   b Find the probability of exactly three arrivals during a randomly chosen one-minute interval.

c Find the probability of at least five arrivals during a randomly chosen five-minute interval.

d Using a suitable approximating distribution, calculate the probability that there are more than 60 arrivals between 6 pm and 7:30 pm.

9 A shopkeeper knows that on average she sells 33 computers a week. She wants to make sure she has enough stock so that for 95% of weeks she has enough computers to meet demand. How many computers should she keep in stock?

10 An undergraduate has noticed that the number of typing errors she makes is, on average, 3 per hour.

a State an assumption that you need to make in order to model the number of typing errors in a randomly chosen hour by a Poisson distribution.

Assume that this model is valid, and that the relevant periods of time are randomly chosen.

b Find the probability that, in one hour, the undergraduate makes more than 2 typing errors.

c Find the probability that, in 15 minutes, the undergraduate makes no typing errors.

d Use a suitable approximation to find the probability that, in six hours, the undergraduate makes more than 20 typing errors.

> **Stop and think**
>
> The number of people in a queue at a supermarket checkout can be modelled by a Poisson distribution.
>
> Suggest an appropriate distribution for the number of people at 5 checkouts. Does the average number of people in a queue at one supermarket checkout make a difference to your answer?

Mathematics in life and work: Group discussion

You are a biologist studying the life forms found in a pond. You find that, on average, there are 5 insect larvae and 0.1 fish in 10 litres of water.

1 Suggest a method you could have used to calculate the average number of larvae and fish.

2 What assumptions are needed to use a Poisson distribution in this context? Are these assumptions valid?

3 The pond has a volume of 6500 litres. What is the probability that there are more than 4000 larvae and fish in total, given that there are 80 fish in the pond?

4 Discuss some scenarios in which you would want to be able to find the probability of the numbers of lifeforms in a pond.

SUMMARY OF KEY POINTS

- A Poisson distribution can be applied when the events:

  - occur independently

  - can only happen singly

  - occur randomly

  - occur at a constant mean rate.

- A random variable X that has a Poisson distribution, with a constant average rate λ, is written as $X \sim Po(\lambda)$.

- The probability of a Poisson distribution is given by $P_r = e^{-\lambda}\dfrac{\lambda^r}{r!}$, where r is the number of the occurrence and λ is the average rate.

- In a Poisson distribution, the mean and variance are equal to λ.

- If random variables X_1, X_2, X_3 ... follow Poisson distributions with mean $\lambda_1, \lambda_2, \lambda_3$ ... respectively, then the random variable, $Y = X_1 + X_2 + X_3 + $ ... also follows a Poisson distribution with mean $\lambda = \lambda_1 + \lambda_2 + \lambda_3 + $ ...

- The binomial distribution $B(n, p)$ may be approximated by $Po(np)$ provided $np < 5$ and $n > 50$.

- The Poisson distribution $Po(\lambda)$ may be approximated by $N(\lambda, \lambda)$ using a continuity correction, provided that $\lambda > 15$.

EXAM-STYLE QUESTIONS

1 The number of chocolate chips in a cookie can be modelled by a Poisson distribution with a mean of 4.5.

   a Calculate the probability that in a randomly chosen cookie there are exactly 5 chocolate chips.

   b Find the probability that in 3 randomly chosen cookies there are 16 chocolate chips.

2 A box contains a large number of sweets. The chance of picking an orange sweet is 8%. A sample of 30 sweets is taken and the number of orange sweets is recorded.

   a Find the probability there are fewer than 3 orange sweets.

   A second sample of 60 sweets is taken and the number of orange sweets in this sample is added to the number of orange sweets in the first sample.

   b Using a Poisson approximation, find the probability there are more than 3 sweets in the combined samples.

3 In a game, two fair, six-sided dice are thrown and the sum of the two scores is recorded. Seventy people play the game. Find, using a suitable approximation, the probability that more than 3 people obtain a score of 12.

4 a The following table shows the probability distribution for the random variable Y.

y	0	1	2	2.25	>2.25
$P(Y = y)$	0.37	0.252	0.15	0.131	0.097

Explain how you can tell from the probability distribution that Y does not have a Poisson distribution.

b The random variable $X \sim Po(\lambda)$. It is given that $P(X = 0) = p$ and $P(X = 2) = 2p$.

i Show that $\lambda = 2$.

ii Find $P(X = 4)$.

5 The average total number of goals scored per football match is 2.3, whenever Sundown United are playing. The goals are scored at random and are independent.

a Find the probability that exactly 4 goals are scored in a randomly chosen match.

It is known that Sundown United score goals at an average rate of 1.4 goals per match.

b Find the probability that a total of 3 goals are scored and Sundown United won the match.

6 The probability that a lightbulb is faulty is 0.005. A random sample of 300 lightbulbs is chosen and the number of faulty lightbulbs, F, is counted.

a Describe fully the distribution of F and also describe a suitable approximating distribution. You must justify your answer.

b Using the approximating distribution, find:

i $P(F > 4)$

ii the smallest value of n such that $P(F = n) < P(F = n - 1)$.

7 A machine breaks down at random at a rate of once every 27 hours.

a Write down two assumptions needed to model this situation as a Poisson distribution.

b Find the probability the machine will work for 20 hours without a breakdown.

c Find the probability that, in a 12-hour shift:

i there are exactly 2 breakdowns

ii there is at least 1 breakdown.

8 A legal administrator makes typing errors at a rate of 1.5 errors for every 500 words written.

a Find the probability that, in a document that is 3000 words long, the administrator makes more than 6 errors.

An important document that is 12 000 words in length needs to be typed. The document must have fewer than 38 mistakes in it, otherwise it must be retyped.

b Find the probability the administrator has to retype the document.

C **9** A new drug is being tested to cure a disease. It is found that it is successful in 97% of patients. It is tested on a random selection of 200 patients. S represents the number of people the drug is successful in.

    a Explain why S cannot be approximated by a Poisson distribution.

    b By considering the number of patients where the drug is not successful, find $P(S \geqslant 195)$.

10 On average, Linh receives 5 emails per day on her personal account. Assuming this can be modelled by a Poisson distribution, find:

    a the probability she gets 7 emails on a randomly chosen day

    b the probability she gets more than 4 emails on a randomly chosen day.

On average, Linh also receives 13 emails per day on her work account.

    c Find the probability that she gets exactly 15 work emails in a day.

    d Find the probability she gets more than 25 emails in total during one day.

PS **11** A steel wire manufacturer knows that 3 faults occur in 10 metres of wire, on average.

    a Find the probability that there are exactly 5 faults in 10 metres of wire.

A shopkeeper buys the wire and sells it in lengths of x metres. The shopkeeper cuts the wire to x metres so that 75% of the pieces of wire have no faults in them.

    b Calculate the length, x, of the pieces.

12 The number of calls received per 10-minute period at a call centre has a Poisson distribution with mean λ, where $\lambda > 50$. If more than 75 calls are received in a 10-minute period, the call centre is unable to meet the demand. It has been found that the probability of being unable to meet demand during a randomly chosen 10-minute period is 0.05.

    a Show that λ satisfies the following equation:

$$\lambda + 1.645\sqrt{\lambda} - 75.5 = 0$$

    b Hence find the probability that there are fewer than 50 calls in a 10-minute period.

13 Cars arrive at the main entrance in a car park at random and at a constant rate of 0.6 cars per minute. Find the probability that:

    a exactly 4 cars arrive in the car park in 5 minutes

    b at least 40 cars arrive in the car park in 1 hour.

The car park has another entrance. The cars arrive at this entrance at random, with an average rate of 0.3 cars per minute.

    c Find the probability that more than 4 cars in total arrive at the car park during a 3-minute interval.

14 A local farm stocks Christmas decorations during the festive season. The sales may be modelled by a Poisson distribution with mean 5.6 per day.

    a Find the probability that, on a particular day, the farm will sell:

        i no decorations

        ii 5 or more decorations.

    b Find the probability that, in a particular week, the farm will sell more than 50 Christmas decorations.

15 David manages a post office in a busy city open seven days a week. The number of recorded letters posted at this office may be modelled by a Poisson distribution with mean 3.2 per day.

    a Find the probability that, at this post office, at least two recorded letters are posted on a particular day.

    b Find the probability that, on any two days in a week, more than six recorded letters are posted.

The town also has another post office managed by Simon. Here the number of recorded letters posted may be modelled by a Poisson distribution with mean 1.8 per day.

    c Find the probability that, on a particular day, the number of recorded letters posted at Simon's post office is fewer than three.

    d The numbers of recorded letters posted at both post offices are independent. Find the probability that, on a particular day, the total number of recorded letters posted at the two post offices is fewer than four.

16 The number, X, of Grade A*s achieved in total by students at a college in their Mathematics examinations each year can be modelled by a Poisson distribution with a mean of 3.

    a Determine the probability that, during a 3-year period, students at the college achieve a total of more than 15 Grade A*s in their Mathematics examinations.

    b The number, Y, of Grade A*s achieved in total by students at a college in their Science examinations each year can be modelled by a Poisson distribution with a mean of 8.

        i Determine the probability that, during a year, students at the college achieve a total of exactly 12 Grade A*s in their Mathematics and Science examinations.

        ii Using a suitable approximation, determine the probability, that during a 5-year period, students at the college achieve a total of more than 45 grade A*s in their Mathematics and Science examinations. Give your answer to 4 significant figures.

17 The number, A, of tablets sold during one day at a computer store in city A can be modelled by a Poisson distribution with mean of 4.5.

The number, B, of tablets sold during one day at a computer store in city B can be modelled by a Poisson distribution with a mean of 6.5.

    a Find the probability that:

        i $P(A = 4)$

        ii $P(B < 5)$

        iii a total of more than 5 tablets is sold from these two stores on one particular day.

b Find the probability that a total of more than 5 tablets are bought from these two stores on at least 4 out of 5 consecutive days.

c The numbers of tablets sold at a computer store in city C, over a period of a week are recorded as:

6   15   8   9   9   10   12

i Calculate the mean and the variance of these data.

ii State, giving a possible reason based on your results in **part i**, whether or not a Poisson distribution provides a suitable model for these data.

18 The number, X, of emails per hour received by a customer service team may be modelled by a Poisson distribution with mean 2.7.

The number, Y, of phone calls per hour received by the same customer service team may also be modelled by a Poisson distribution with mean 5.6.

a For any particular hour, find:

i $P(X = 4)$

ii $P(5 < Y \leqslant 8)$.

b **i** Write down the distribution of T, the total number of emails and phone calls per hour received by the customer service team.

ii Determine $P(8 \leqslant T \leqslant 10)$.

iii Hence find the probability that the customer service team receive a total of at least 8 but at most 10 emails and phone calls over three consecutive hours.

c Using a suitable approximation, find the probability that the customer service team receive fewer than 50 emails and phone calls in a working day. You can assume that a working day is from 9am to 5pm.

19 A year 7 student is proud of her times table skill. The statistics shows that there is 0.6% inaccuracy in her tests. Each test has 100 questions and is practised daily by the student.

a Suggest, with a reason, an appropriate distribution to model the number of correct answers in a test at random.

b Give two conditions that need to be satisfied for this distribution to be a suitable model.

c Using a normal approximation, find the probability that a test contains more than one mistake.

20 The number of vehicles arriving at a tunnel entrance during a 5-minute period can be modelled by a Poisson distribution with mean 3.6.

a State the value for the standard deviation of the number of vehicles arriving at the tunnel during a 10-minutes period.

b Find the probability that no vehicles arrive in a 10-minute period.

c Find:

   i the probability that the number of vehicles arriving in a 5-minute period is within one standard deviation

   ii the probability that at least three vehicles arrive in each of three successive 5-minute periods.

d Using a suitable approximation, find the probability that more than 25 vehicles arrive in an hour.

Mathematics in life and work

A biologist is studying the habits of male snow leopards and has set up a camera to film in a particular location. It is known that snow leopards are solitary creatures, so always appear alone.

1 State another condition for the number of leopards captured on camera in a day to have a Poisson distribution.

It is given that the number of leopards filmed per day has the distribution Po(0.8).
Find the probability that:

2 2 leopards are filmed on a particular day

3 at least 4 leopards are filmed in a week

4 5 leopards are seen in a week, if no leopards are seen on the first day of the week.

2 LINEAR COMBINATIONS OF RANDOM VARIABLES

Mathematics in life and work

In this chapter you will learn how to find the mean and variance of the linear function of a random variable and how to combine multiple random variables, using a linear combination. You will learn how to find probabilities, using linear combinations of Poisson and normal distributions.

The modelling of situations as a combination of random variables has a wide range of applications in many different careers, for example:

> If you were a carpenter you would be able to use a linear combination of normal distributions to model how thick a piece of laminated wood will be.

> If you worked for an airline you would be able to work out the expected mass of the luggage on a flight.

> If you were a lift manufacturer you would be able to calculate the probability of a certain number of people weighing more than the safety limit of the lift.

In this chapter you will focus on the use of random variables in manufacturing.

LEARNING OBJECTIVES

You will learn how to:

> solve problems using the results:

$\mathrm{E}(aX + b) = a\mathrm{E}(X) + b$

$\mathrm{Var}(aX + b) = a^2\mathrm{Var}(X)$

$\mathrm{E}(aX + bY) = a\mathrm{E}(X) + b\mathrm{E}(Y)$

$\mathrm{Var}(aX + bY) = a^2\mathrm{Var}(X) + b^2\mathrm{Var}(Y)$ for independent X and Y

> define the distribution of $X + Y$, if X and Y have independent Poisson distributions

> find probabilities with the distribution $aX + b$, if X has a normal distribution

> define a distribution of $aX + bY$, if X and Y both are independently normal distributed.

LANGUAGE OF MATHEMATICS

Key words and phrases you will meet in this chapter:

> combination, expectation, linear, variance

PREREQUISITE KNOWLEDGE

You should already know how to:

> find the mean and variance of a random variable

> calculate probabilities from a Poisson distribution

> calculate probabilities from a normal distribution.

You should be able to complete the following questions correctly:

1   Find the expectation and variance of the random variable with the distribution shown below.

x	−1	0	1	2
P($X = x$)	0.3	0.25	0.15	0.3

2   Y is a random variable that follows the Poisson distribution with a mean of 5.

   a Write down the variance of Y.

   b Find P($Y = 4$).

   c Find P($Y > 3$).

3   The height of a group of men is normally distributed with mean of 175.5 cm and standard deviation of 7.37 cm. Find the probability that:

   a a randomly chosen man has height less than 190 cm

   b a randomly chosen man has height between 170 cm and 185 cm.

2.1 Expectation and variance of a linear function of random variables

In **Probability & Statistics 1, Chapter 3, Discrete random variables** you learned about expectation and variance. This can be extended to a linear function of the random variable.

Let X be a random variable with the following probability distribution.

x	0	1	2	3	4
P($X = x$)	0.1	0.2	0.5	0.1	0.1

You can find the expectation and variance of X.

$$E(X) = 0 \times 0.1 + 1 \times 0.2 + 2 \times 0.5 + 3 \times 0.1 + 4 \times 0.1$$

$$= 1.9$$

$E(X) = \mu = \Sigma \, xP(X = x)$

$$Var(X) = 0^2 \times 0.1 + 1^2 \times 0.2 + 2^2 \times 0.5 + 3^2 \times 0.1 + 4^2 \times 0.1 - 1.9^2$$

$$= 1.09$$

Y is a random variable where $Y = 2X - 3$. The distribution of Y is:

X	0	1	2	3	4
Y	−3	−1	1	3	5
$P(X = x)$	0.1	0.2	0.5	0.1	0.1

For $X = 0$ the corresponding Y value is $Y = 2(0) - 3 = -3$.

Note that the probability for Y is the same as the corresponding X value.

$E(Y) = -3 \times 0.1 + -1 \times 0.2 + 1 \times 0.5 + 3 \times 0.1 + 5 \times 0.1$

$\quad = 0.8$

$Var(Y) = (-3)^2 \times 0.1 + (-1)^2 \times 0.2 + 1^2 \times 0.5 + 3^2 \times 0.1 + 5^2 \times 0.1 - 0.8^2$

$\quad = 4.36$

This example shows that the connection between $E(X)$ and $E(Y)$ is $E(Y) = 2E(X) - 3$. The variance of X and Y are also connected by $Var(Y) = 2^2Var(X)$. This is an example of the general results:

$E(aX + b) = aE(X) + b$

$Var(aX + b) = a^2Var(X)$

KEY INFORMATION

The expectation of a linear function $aX + b$ is
$E(aX + b) = aE(X) + b$

The variation of a linear function $aX + b$ is
$Var(aX + b) = a^2Var(X)$

Stop and think Use a real-life example, such as the height of people, to illustrate why adding a constant to a random variable doesn't affect the variance.

Example 1

The random variable X has a mean of 5 and a variance of 3. Find the mean and variance of:

a $W = 3X + 2$ **b** $Y = 10 - 3X$.

Solution

a $E(W) = 3 \times E(X) + 2$

$\quad = 3 \times 5 + 2$

$\quad = 17$

$Var(W) = 3^2 \times Var(X)$

$\quad = 9 \times 3$

$\quad = 27$

b $E(Y) = -3E(X) + 10$

$\quad = -3 \times 5 + 10$

$\quad = -5$

$Var(Y) = (-3)^2 \times Var(X)$

$\quad = 9 \times 3$

$\quad = 27$

Expectation and variance of linear combinations of random variables

You may know the mean and variance of two individual random variables and want to know the mean and variance of the combination of these random variables. For example, you may want the mean and variance of the total mass of a jar of jam, given the mean and variance of the mass of the jar and the mass of the jam separately.

Let X and Y be random variables with the following probability distributions:

x	0	1	2
$P(X = x)$	0.3	0.5	0.2

y	1	2
$P(Y = y)$	0.7	0.3

The mean of X is $\mu = E(X) = 0.9$ and variance of X is $\sigma^2 = \text{Var}(X) = 0.49$. The mean of Y is 1.3 and variance of Y is 0.21.

Let T be the linear combination, $2X + 3Y$. The possible values for T are as follows.

		Values of X		
		0	1	2
Values of Y	1	3	5	7
	2	6	8	10

The probability distribution table of T is:

t	3	5	6	7	8	10
$P(T = t)$	0.21	0.35	0.09	0.14	0.15	0.06

$E(T) = 3 \times 0.21 + 5 \times 0.35 + 6 \times 0.09 + 7 \times 0.14 + 8 \times 0.15 + 10 \times 0.06$

$\quad = 5.7$

$\text{Var}(T) = 3^2 \times 0.21 + 5^2 \times 0.35 + 6^2 \times 0.09 + 7^2 \times 0.14 + 8^2 \times 0.15$
$\qquad\qquad + 10^2 \times 0.06 - 5.7^2$

$\quad = 3.85$

$E(T) = 5.7 = 2 \times 0.9 + 3 \times 1.3 = 2 \times E(X) + 3 \times E(Y)$

$\text{Var}(T) = 3.85 = 2^2 \times 0.49 + 3^2 \times 0.21 = 2^2\text{Var}(X) + 3^2\text{Var}(Y)$

This leads to two general formulae:

$E(aX + bY) = aE(X) + bE(Y)$

$\text{Var}(aX + bY) = a^2\text{Var}(X) + b^2\text{Var}(Y)$

KEY INFORMATION

The expectation of a linear combination $aX + bY$ is
$E(aX + bY) = aE(X) + bE(Y)$

The variance of a linear combination $aX + bY$
is $\text{Var}(aX + bY)$
$= a^2\text{Var}(X) + b^2\text{Var}(Y)$

Example 2

The random variable X has mean 3 and variance 2.

The random variable Y has mean 10 and variance 1.

Find the mean and variance of:

a $X + Y$ **b** $3X + 2Y$ **c** $Y - 2X$.

Solution

a $E(X + Y) = E(X) + E(Y)$

$\qquad = 3 + 10$

$\qquad = 13$

$Var(X + Y) = Var(X) + Var(Y)$

$\qquad = 2 + 1$

$\qquad = 3$

b $E(3X + 2Y) = 3E(X) + 2E(Y)$

$\qquad = 3 \times 3 + 2 \times 10$

$\qquad = 29$

$Var(3X + 2Y) = 3^2 Var(X) + 2^2 Var(Y)$

$\qquad = 3^2 \times 2 + 2^2 \times 1$

$\qquad = 22$

c $E(Y - 2X) = E(Y) + (-2)E(X)$

$\qquad = 10 - 2 \times 3$

$\qquad = 4$

$Var(Y - 2X) = Var(Y) + (-2)^2 Var(X)$

$\qquad = 1 + 4 \times 2$

$\qquad = 9$

> Note that when you square the coefficient of X you get a positive number. Therefore, when you are calculating the variance you never subtract, despite there being a subtraction in the combination.

Stop and think A real-life example of a linear combination of random variables provided earlier in the chapter was that of the combined mass of a jar and the jam inside it. What other real-life examples can you think of?

Expectation and variance of more than one observation of a random variable

If a random variable L represents the length of a fence panel, you may want to find the mean and variance of five fence panels.

You are taking five independent observations of L, as you would have five different fence panels.

Calculate the expectation using the formula for the expectation of a linear combination of random variables:

$$E(L_1 + L_2 + L_3 + L_4 + L_5) = E(L_1) + E(L_2) + E(L_3) + E(L_4) + E(L_5)$$
$$= 5E(L)$$

The variance of five fence panels would be:

$$Var(L_1 + L_2 + L_3 + L_4 + L_5) = Var(L_1) + Var(L_2) + Var(L_3) + Var(L_4)$$
$$+ Var(L_5)$$
$$= 5Var(L)$$

This is not the same as if you wanted to find the variance of $5L$ which would be $5^2 \times Var(L)$. In this case, you are taking one observation each time and repeating it five times. The spread of possible observations increases 5 times.

Example 3

Apples are sold in packets of four. Each apple has a mean mass of 60 grams and variance of 10. The packaging has a mean mass of 30 grams and variance of 3. Find the mean and variance of:

a the mass of the four apples

b the total mass of the packet of apples.

Solution

a $E(4 \text{ apples}) = E(A_1 + A_2 + A_3 + A_4)$

$$= 4 E(A)$$
$$= 4 \times 60$$
$$= 240 \text{ grams}$$

$Var(4 \text{ apples}) = Var(A_1 + A_2 + A_3 + A_4)$

$$= 4Var(A)$$
$$= 4 \times 10$$
$$= 40 \text{ grams}$$

There are four different apples being put in the packet so you need to consider four independent observations of the random variable A.

b $E(4 \text{ apples} + \text{packaging}) = E(A_1 + A_2 + A_3 + A_4 + P)$

$$= 4E(A) + E(P)$$
$$= 4 \times 60 + 30$$
$$= 270 \text{ grams}$$

$Var(4 \text{ apples} + \text{packaging}) = Var(A_1 + A_2 + A_3 + A_4 + P)$

$$= 4Var(A) + Var(P)$$
$$= 4 \times 10 + 3$$
$$= 43 \text{ grams}$$

Exercise 2.1A

1 The random variable X has the probability distribution shown below.

x	0	1	2	3
$P(X = x)$	0.1	0.4	0.3	0.2

  a Find $E(X)$ and $Var(X)$.

  b Y is a variable defined by $Y = 2X - 1$.

    i Use the general formula to find $E(Y)$ and $Var(Y)$.

    ii Verify your answers, using the probability distribution of Y.

2 A random variable, X, has a mean of 13 and variance of 4. Find the expectation and variance of:

  a $M = 2X$ **b** $P = X - 6$ **c** $D = 15 - X$.

3 The random variables X and Y have means of 5 and 6 and variances of 1 and 3 respectively. Find the mean and standard deviation of:

  a $X + Y$ **b** $3X - Y$ **c** $\frac{1}{2}X - 3Y$.

4 The random variable X has mean 5 and variance 2. Find the mean and standard deviation of three independent observations of X.

(MM) 5 The random variable T is the score on a fair tetrahedral die with sides numbered 1, 2, 3 and 4 and H is the number of heads when a fair coin is flipped once. Find the expectation and variance of:

  a T **b** $3H$ **c** $T - 3H$.

(C) 6 There are two variables, X and Y, with Poisson distributions and mean of 6 and 14 respectively.

  a Find the mean and variance of:

    i $X - Y$ **ii** $3X + 4$.

  b Decide whether the distributions in part **a** are Poisson distributions. Give a reason for your answer.

(PS) 7 B and C are random variables where $C = 3B + 2$. The expectation of C is 8 and the variance is 8. Calculate $E(B)$ and $Var(B)$.

8 A chocolate bar has three layers: a milk chocolate layer, a white chocolate layer topped off with another milk chocolate layer. The depth of the milk chocolate layers is on average 2 mm and has a variance of 0.15. The white chocolate layer has a mean depth of 1.3 mm with a variance of 0.1. Find the mean and variance of the depth of the chocolate bar.

(MM) (PS) 9 Four sweets are taken from a bag that contains a mixture of lemon and strawberry flavours. The mean and variance of the number of lemon sweets taken are 3 and 0.75 respectively. Find the mean and variance of the number of strawberry sweets taken.

(PS) 10 The random variables X and Y have means of 5 and 2, and variances of 1 and 0.5 respectively. The expectation of $aX + bY$ is 17 and the variance of $aX + bY$ is 9.5. Find the possible values of a and b.

Mathematics in life and work: Group discussion

You are working for an airline and you have been asked to determine the best booking policy. The airline wants to ensure as many seats as possible are filled so they sell more tickets than there are seats, as they know that some passengers do not turn up for their flight. From past experience the number, E, of economy class passengers and the number, B, of business class passenger who miss a flight have the distributions shown below.

e	0	1	2	3
$P(E = e)$	0.2	0.5	0.25	0.05

b	0	1	2	3
$P(B = b)$	0.3	0.5	0.15	0.05

1 Find the expectation and variance of E and B.

2 Find the expectation and variance of the total number of passengers who miss a flight.

The company has a compensation policy for people who cannot fly due to the flight being overbooked. An economy passenger receives \$250 and a business class passenger receives \$550.

3 By how many of each type of passenger would you recommend the airline to overbook the flight?

2.2 Expectation and variance of the sum of independent Poisson distributions

In **Chapter 1 The Poisson distribution**, you saw that the mean and variance of a Poisson distribution are equal to each other. This means that you can decide whether the sum of Poisson distributions also follows a Poisson distribution by calculating the mean and variance. In general, you can say that if X and Y have independent Poisson distributions, then $X + Y$ has a Poisson distribution. This can be demonstrated.

KEY INFORMATION

If X and Y have independent Poisson distributions, then $X + Y$ has a Poisson distribution with
$E(X + Y) = E(X) + E(Y)$ and
$Var(X + Y) = Var(X) + Var(Y)$

Let $X \sim Po(3)$ and $Y \sim Po(8)$.

$E(X + Y) = E(X) + E(Y)$

$\qquad = 3 + 8$

$\qquad = 11$

$Var(X + Y) = Var(X) + Var(Y)$

$\qquad = 3 + 8$

$\qquad = 11$

Since the expectation and variance of $X + Y$ are equal, then $X + Y \sim Po(11)$.

Example 4

Cars and lorries pass a junction on a road at an average rate of 15 cars and 3 lorries every 10 minutes. Assuming there are no other types of vehicle, find the probability that:

a exactly 10 vehicles pass in a 10-minute period

b more than 3 vehicles pass the junction in a 5-minute period.

Solution

a Let C be the number of cars in 10 minutes, where $C \sim \text{Po}(15)$.

Let L be the number of lorries in 10 minutes, where $L \sim \text{Po}(3)$.

Let T be the total number of vehicles in 10 minutes, where $T = C + L$.

$E(T) = E(C) + E(L) = 15 + 3 = 18$ so $T \sim \text{Po}(18)$.

$$P(T = 10) = e^{-18} \frac{18^{10}}{10!}$$
$$= 0.0150 \ (3 \text{ s.f.})$$

b Let U be the number of vehicles in a 5 minutes, so $U \sim \text{Po}(9)$.

$$P(U > 3) = 1 - P(U \leq 3)$$
$$= 1 - e^{-9}\left(\frac{9^0}{0!} + \frac{9^1}{1!} + \frac{9^2}{2!} + \frac{9^3}{3!} \right)$$
$$= 0.979 \ (3 \text{ s.f.})$$

Example 5

A company makes two types of light bulb. Light bulb A has a failure rate of 2% and light bulb B has a failure rate of 3.5%.

a A random sample of 60 of each type of light bulb is taken. Find the probability, using a suitable approximation, that there are at least 4 light bulbs of either type that fail.

b A random sample of n of each type of bulb is taken. Find the sample size needed to ensure the probability that at least one light bulb of either type fails is more than 0.95.

Solution

a Let A be the number of type A light bulbs that fail, so $A \sim \text{B}(60, 0.02)$.

$np = 60 \times 0.02 = 1.2 < 5$

Since $n > 50$ and $np < 5$ a Poisson approximation can be used with $A \sim \text{Po}(1.2)$.

Let B be the number of type B light bulbs that fail, so $B \sim \text{B}(60, 0.035)$.

$np = 60 \times 0.035 = 2.1 < 5$

Since $n > 50$ and $np < 5$ a Poisson approximation can be used with $B \sim \text{Po}(2.1)$.

Let T be the total number of light bulbs that fail, where $T = A + B$.

$T \sim \text{Po}(3.3)$

$P(T \geqslant 4) = 1 - P(T < 4)$

$$= 1 - e^{-3.3}\left(\frac{3.3^0}{0!} + \frac{3.3^1}{1!} + \frac{3.3^2}{2!} + \frac{3.3^3}{3!} \right)$$

$$= 0.420 \text{ (3 s.f.)}$$

b To use a Poisson approximation, you must assume that $n > 50$. You also require $np < 5$, so:
for type A, $0.02n < 5 \quad \Rightarrow n < 250$

for type B, $0.035n < 5 \quad \Rightarrow n < 142.9$

n must be an integer, so for the solution to be valid $51 \leqslant n \leqslant 142$

$E(T) = 0.02n + 0.035n = 0.055n$

Therefore $T \sim \text{Po}(0.055n)$

$P(T \geqslant 1) > 0.95$

$1 - P(T = 0) > 0.95$

$P(T = 0) < 0.05$

$e^{-0.055n} \times \dfrac{(0.055n)^0}{0!} < 0.05$

$e^{-0.055n} < 0.05$

$-0.055n < \ln 0.05$

$n > \dfrac{\ln 0.05}{-0.055}$

$n > 54.468$

The sample size should contain at least 55 but no more than 142 light bulbs of each type.

> Since $(0.055n)^0 = 1$ and $0! = 1$ then $\dfrac{(0.055n)^0}{0!} = 1$

> Take ln of each side.

> The inequality changes direction because you are dividing by a negative number.

Exercise 2.2A

1 X and Y are independent random variables where $X \sim \text{Po}(4)$ and $Y \sim \text{Po}(2)$. Find:

   a $P(X + Y = 5)$ **b** $P(X + Y < 3)$ **c** $P(X + Y > 5)$.

2   In a large field there are exactly 2 types of flower, which grow independently. There are buttercups that occur at a rate of 3.6 flowers per square metre and dandelions that occur at a rate of 0.9 flowers per square metre. Find the probability that there are between 8 and 12 flowers (inclusive) in a randomly chosen 2 square metre section of the field.

3   The emissions from two samples of radioactive material are measured. The first sample emits particles at a rate of 3.4 per minute. The second sample emits particles at a rate of 11 particles in 5 minutes.

   a   Find the probability that there are 6 particles emitted by the second sample in a randomly chosen 2-minute period.

   b   Find the probability that there is a total of fewer than 5 particles emitted by the two samples in a randomly chosen minute.

4   On average, at a fast food restaurant, 8 orders are placed every 10 minutes during opening hours.

   a   State two conditions that must be satisfied for the number of orders placed to follow a Poisson distribution.

   b   Assuming these conditions are met, find the probability that fewer than 5 orders are placed in any randomly chosen 10-minute period.

The restaurant also has a drive-through booth, where customers can order without leaving their cars. On average, 4 cars order at the drive-through booth in a 30-minute period.

   c   Calculate the probability that exactly 15 orders in total are placed in a randomly chosen 30-minute period.

5   In a household goods store, a salesperson sells on average 6 fridges and 5 cookers per week. The numbers of fridges and cookers sold follow a Poisson distribution and are independent of each other.

   a   Find the mean and variance for the total number of fridges and cookers sold in a week.

   b   Find the probability that the salesperson sold a total of 8 fridges or cookers in a week.

The store managers offer a bonus to anyone who is able to sell more than a total of 55 fridges or cookers in a month, assuming four weeks in a month.

   c   Find the probability that the salesperson gets a bonus in a randomly chosen month.

6   A gardener plants two types of seeds, A and B. The probability that a randomly chosen seed of type A does not grow is 1.5% and the corresponding probability for a seed of type B is 0.3%. The gardener plants 55 of each type of seed.

   a   Find, using a suitable approximation, the probability that exactly 106 plants grow.

   b   What is the maximum number of each type of seed that can be planted if the gardener wants the probability that all the seeds grow to be at least 5%? You may assume that an equal number of each type are planted.

7 A museum offers two different types of entrance ticket.

Ticket A offers a free drink and ticket B offers a free souvenir. Based on past experience, the ticket office sells 25 of ticket A and 30 of ticket B in 5 hours on average. Using a normal approximation, the probability that:

a exactly 38 type A tickets are sold in 6 hours

b more than 5 tickets are sold in half an hour.

8 The numbers of phone calls and emails to a university office are monitored. During working hours in September the number of calls received follows a Poisson distribution with mean 2.6 per 10-minute interval. The number of emails received also follows a Poisson distribution with mean 4.8 per 10-minute interval.

a Find the probability that two or fewer people contacted the office during a particular 15-minute interval.

b Find the greatest number of emails and calls that the university office receives in an hour, with a probability of greater than 0.01.

9 Niamh runs a local post office. The numbers of first class and second class letters posted at this office may be modelled by Poisson distribution with mean 7.6 and 4.8 per day respectively.

Find the probability that at this post office:

a two or fewer letters are posted in half a day

b a total of four or more letters are posted on three consecutive days.

10 A factory makes two types of battery. Battery A has a failure rate of 1.4% and battery B has a failure rate of 2.6%. A random sample of 100 of each type of battery is taken.

Using a suitable approximation, find the probability that there are at least 5 batteries of either type that are faulty.

Mathematics in life and work: Group discussion

You are the manager of a maternity hospital with responsibility for ensuring there are sufficient numbers of staff on duty during each shift. The shift timings and the average proportion of babies born during each shift are as follows:

Shift A: 9 am to 5 pm (45%)

Shift B: 5 pm to 1 am (32.5%)

Shift C: 1 am to 9 am (22.5%)

The average number of births per year at the hospital is 3200.

1 What is the expected number of deliveries for each shift on a typical day?

2 What is the probability that more than 3 babies are born during every shift in a given day?

3 What is the probability that at least 10 babies in total are born during a given day?

4 Given that twice as many babies are born during shift A compared to shift C, is it likely that shift A will have twice as many staff as shift C?

2.3 Linear combinations of normal distributions

In **Mathematics: Probability & Statistics, 1 Chapter 4,** you saw that the normal distribution can be used to model a continuous random variable. Visualising the normal distribution, it is clear that any translation or stretch parallel to x or y will still result in a bell shape. This shows that a linear function of a normally distributed random variable is itself a normal distribution. In other words, if X has a normal distribution, then so does $aX + b$.

This result also holds for linear combinations of normally distributed random variables. In other words, if X and Y have independent normal distributions then $aX + bY$ has a normal distribution.

The results in Section 2.1 can be used to calculate the mean and variance of the resulting normal distributions.

If X is normally distributed with mean μ and variance σ^2, then $aX + b \sim \mathrm{N}(a\mu + b, a^2 \sigma^2)$.

If X is normally distributed with mean μ_X and variance σ_X^2 and Y is normally distributed with mean μ_Y and variance σ_Y^2, then $aX + bY \sim \mathrm{N}(a\mu_X + b\mu_Y, a^2\sigma_X^2 + b^2\sigma_Y^2)$.

> **KEY INFORMATION**
>
> If X has a normal distribution, then $aX + b$ has a normal distribution.

> **KEY INFORMATION**
>
> If X and Y have independent normal distributions then $aX + bY$ has a normal distribution.

> From Section 2.1, the mean linear function of a random variable is
> $\mathrm{E}(aX + b) = a\mathrm{E}(X) + b$ and $\mathrm{Var}(aX + b) = a^2\mathrm{Var}(X)$

> From Section 2.1, the mean of linear function of a random variable is
> $\mathrm{E}(aX + bY) = a\mathrm{E}(X) + b\mathrm{E}(Y)$ and $\mathrm{Var}(aX + bY)$
> $= a^2\mathrm{Var}(X) + b^2\mathrm{Var}(Y)$

Example 6

A car rental service charges $25 to hire a car plus 20 cents per kilometre driven. The distance that Bilal drives when he rents a car is normally distributed with mean of 300 km and variance of 40.

a Find the probability of Bilal spending more than $88 on his car rental.

Selina rents a car from a different company that charges a $40 hire fee plus 15 cents per kilometre driven. Selina drives on average 270 km with a variance of 70.

b Find the probability that Bilal pays more than Selina.

Solution

a Let B be the number of kilometres Bilal drives, so $B \sim \mathrm{N}(300, 40)$.

Let C be the total cost of hiring a car $C = 25 + 0.2B$

$\mathrm{E}(C) = 25 + 0.2 \times \mathrm{E}(B) = 25 + 0.2(300) = 85$

$\mathrm{Var}(C) = 0.2^2 \times \mathrm{Var}(B) = 0.2^2(40) = 1.6$

So $C \sim (85, 1.6)$

$P(C > 88) = P\left(Z > \dfrac{(88 - 85)}{\sqrt{1.6}} \right)$

$= P(Z > 2.372)$

$= 1 - P(Z < 2.372)$

$= 1 - 0.9912$

$= 0.0088$

b Let S be the number of kilometres Selina drives, so
$S \sim \mathrm{N}(270, 70)$

Let T be the total cost of Selina hiring a car $T = 40 + 0.15S$

$\mathrm{E}(T) = 40 + 0.15 \times \mathrm{E}(S) = 40 + 0.15(270) = 80.5$

$\mathrm{Var}(T) = 0.15^2 \times \mathrm{Var}(S) = 0.15^2(70) = 1.575$

So $T \sim (80.5, 1.575)$

For Bilal to pay more than Selina, then $C > T$ must be true. This inequality can be rearranged to give $C - T > 0$, which means the problem can be solved using the distribution of $C - T$.

$\mathrm{E}(C - T) = \mathrm{E}(C) - \mathrm{E}(T) = 85 - 80.5 = 4.5$

$\mathrm{Var}(C - T) = 1^2 \times \mathrm{Var}(C) + (-1)^2 \times \mathrm{Var}(T) = 1.6 + 1.575 = 3.175$

So $C - T \sim \mathrm{N}(4.5, 3.175)$

$\mathrm{P}(C > T) = \mathrm{P}(C - T > 0)$

$$= \mathrm{P}\left(Z > \frac{(0 - 4.5)}{\sqrt{3.175}} \right)$$

$$= \mathrm{P}(Z > -2.525)$$

$$= \mathrm{P}(Z < 2.525)$$

$$= 0.994$$

> Using the symmetry of the normal distribution, $\mathrm{P}(Z > -z) = \mathrm{P}(Z < z)$.

Example 7

A lift has a mass of 230 kg and can hold 10 adults. The mass of adults is normally distributed with mean 68 kg and variance of 900. The lift is safe if the total mass is less than 1150 kg.
Find the probability the total mass of the lift and 10 adults is below the safety limit.

Solution

Let M be the mass of an adult where $M \sim \mathrm{N}(68, 900)$.

Let T be the total mass of the lift and 10 adults.

$T = 230 + M_1 + M_2 + M_3 + M_4 + M_5 + M_6 + M_7 + M_8 + M_9 + M_{10}$

$\mathrm{E}(T) = 230 + 10(68) = 910$

$\mathrm{Var}(T) = 10(900) = 9000$

So $T \sim \mathrm{N}(910, 9000)$

$\mathrm{P}(T < 1150) = \mathrm{P}\left(Z < \frac{1150 - 910}{\sqrt{9000}} \right)$

$$= \mathrm{P}(Z < 2.530)$$

$$= 0.994$$

> Since the masses of the adults are independent observations this cannot be written as $230 + 10M$.

Exercise 2.3A

1. X and Y are independent random variables, where $X \sim N(40, 9)$ and $Y \sim N(20, 4)$.

   a State the distribution of $X + Y$.

   b Find the mean and variance of $X + Y$.

   c Find $P(X + Y < 65)$.

   d Find $P(X + Y > 50)$.

2. A baker makes cupcakes and sells them in packs of four. She knows that cupcakes have a mean mass of 95 g and variance 13.5 and that their masses follow a normal distribution. The packaging weighs 60 g.

   a Find the distribution of the mass of a pack of 4 cupcakes.

   b Find the probability that a pack weighs:

       i more than 450 g   **ii** between 420 g and 430 g.

3. In a town, the heights of the men are normally distributed with mean 174 cm and variance 28. The heights of women are normally distributed with mean 165 cm and variance 20. A man and woman are chosen at random. Find the probability that:

   a the woman is taller than the man

   b the man is more than 5 cm taller than the woman

   c there is more than 10 cm difference between the heights of the two people.

4. A maths test has two papers. The marks for each paper are normally distributed. Paper 1 has a mean of 60 and standard deviation of 8 and Paper 2 has a mean of 100 and standard deviation 20. The marks for both papers are added together to give the overall grade.

   a Find the probability that a randomly selected student scores less than 150.

   b The top 20% of students get an A. What is the lowest mark that will result in an A grade?

5. The diameters of bolts are normally distributed with mean 2.3 cm and standard deviation 0.02. The diameters of the holes in the corresponding nuts are also normally distributed with mean 2.4 cm and standard deviation 0.03.

   a Find the probability that a randomly chosen bolt does not fit into a randomly chosen nut.

   A random sample of five pairs of nuts and bolts is taken.

   b Find the probability that at least one pair do not fit.

6. Johan and Thambo each time how long they take to solve the crossword in a daily newspaper. They know from past crosswords that their times are normally distributed with means of 15 and 22 minutes and variances 8 and 2.8 respectively. Find the probability that, on a randomly chosen day:

   a Johan completes the crossword faster than Thambo

   b Johan completes the crossword in less than half the time that it takes Thambo.

PS **7** The random variables X and Y have the following distributions:

$X \sim N(10, \sigma^2)$ and $Y \sim N(\mu, 15)$ and $3Y - 2X \sim N(0, 278)$

Find μ and σ.

8 Tori competes in triathlon races comprised of a run, a swim and a bike ride. From her past races she knows the times taken for each activity are normally distributed and can be summarised as:

	Run	Swim	Bike
Mean (minutes)	61	37	88
Variance	9	5	12

a Find the probability that Tori finishes the triathlon in under 3 hours.

b Given that Tori completes the swim section in exactly 35 minutes, find the probability that she finishes the triathlon in under 2 hours and 55 minutes.

9 A drinks crate holds six bottles. The mass of the bottles is normally distributed with mean 165 grams and standard deviation 4.5 grams. The mass of the drinks crate is also normally distributed with mean 25 grams and standard deviation 2 grams.

a Find the probability that the total mass of the crate and the bottles is less than 1 kg.

If the total mass of the crate and bottles is more than 1.02 kg, $5 will be charged for delivery.

b Calculate the expected delivery charge.

10 A mobile phone company has two different monthly Pay As You Go tariffs.

Hannah uses Tariff A, which charges $10 for 3 GB of data, then $2 for each additional 0.1 GB of data.

The average amount of data Hannah uses each month is 3.5 GB of data with variance 1.6.

a Find the probability of Hannah spending less than $15 on Tariff A next month.

Lucy uses Tariff B, which charges $15 for 5 GB of data, then $2.50 for each 0.1 GB of data. Lucy uses an average 4 GB of data with variance 1.2 each month.

b Find the probability that Hannah pays more than Lucy.

SUMMARY OF KEY POINTS

> The expectation and variance of a linear function of a random variable X is given by
> $E(aX + b) = aE(X) + b$ and $Var(aX + b) = a^2 Var(X)$.

> The expectation and variance of a linear combination of random variables X and Y is given by
> $E(aX + bY) = aE(X) + bE(Y)$ and $Var(aX + bY) = a^2 Var(X) + b^2 Var(Y)$.

> If X and Y have independent Poisson distributions, then $X + Y$ has a Poisson distribution.

> If X has a normal distribution, then $aX + b$ has a normal distribution.

> If X and Y have independent normal distributions then $aX + bY$ has a normal distribution.

EXAM-STYLE QUESTIONS

1. X is a random variable that can take the values -1, 0 and 3 with probability p, $2p$ and $3p$ respectively.

   a Find the probability distribution of X.

   b Calculate the expectation and variance of X.

   Y is a random variable defined by $Y = X_1 + X_2 + X_3$ where X_1, X_2 and X_3 are independent observations of X.

   c Find $E(Y)$ and $Var(Y)$.

2. Pens are sold in packets of five. The probability distribution for the number of faulty pens in a pack is shown below.

f	0	1	2	3	4
$P(F = f)$	0.7	0.2	0.05	0.04	0.01

   a Find the expectation and variance of F.

   Bao buys five packets of pens, chosen at random.

   b Find the expectation and variance of the number of faulty pens in five packets, $F_1, F_2, ..., F_5$.

3. A science exam is based on two components: a practical experiment and a written test. The test is worth $\frac{2}{3}$ of the final grade and the experiment is worth $\frac{1}{3}$. The experiment marks follow a normal distribution with mean 35 and variance 30 and the marks on the written test also follow a normal distribution with mean 52 and variance 45.

   a Find the mean and standard deviation of the combined mark, taking into account how much each part is worth.

   b Find the probability that a randomly chosen person's combined score is above 55.

4. The random variable X has a mean of 2 and variance 3. It is given that $E(aX + b) = 10$ and $Var(aX + b) = 27$. Find the possible values of a and b.

5 X and Y are Poisson distributions with mean 6.4 and 3.8 respectively. Calculate:

a $P(X + Y = 10)$

b $P(X + Y > 5)$

c Z is defined as $Z \sim \text{Po}(5.9)$. Calculate $P(X + Y + Z \leqslant 18)$.

6 A manufacturer advertises that there is 300 g of honey in every jar. He knows that the mean amount of honey in a jar is 302 g and that 90% of jars must contain at least 300 g.

a Calculate the variance of the amount of honey in a jar.

The mass of a jar follows a normal distribution with mean 150 g and standard deviation 10. The lid of the jar is a constant mass of 33 g.

b Find the probability a full jar with a lid weighs between 480 g and 500 g.

c Find the probability that the total mass of three full jars, including their lids, exceeds 1.5 kg.

7 A shopkeeper counts how many customers, S, use the self-checkout machines in a supermarket in 15 minutes. She finds that there are 7 customers every 15 minutes on average.

a State two conditions needed for a Poisson distribution to be used.

The shopkeeper also finds that the number of customers who use a staffed checkout is 3 customers every 10 minutes.

b Find the probability that a total of 9 customers attend a checkout in any 15-minute period.

c Find the probability that between 20 and 25 (inclusive) customers attend a checkout in any 30-minute period.

8 A drinks machine sells both black and white coffee. For both types of drink the machine dispenses, on average, 150 ml of coffee with a variance of 25. The white coffee is made by adding in milk. The quantity of milk dispensed follows a normal distribution with mean 12 ml and variance 6. The cup manufacturer wants to minimise costs so decides to make the smallest size possible such that 95% of all the white coffee drinks dispensed will fit into the cup. What capacity should the cups be?

9 A supermarket sells cereal in different sized boxes. Small boxes have a mean mass of 500 g and variance 80. Large boxes have a mean weight of 750 g with variance of 110.

a Find the probability that two randomly chosen small boxes weigh less than 0.98 kg.

b Find the probability that the combined weight of three small boxes exceeds the combined weight of two large boxes, where all selections are made randomly.

10 A point on a train line has two tracks. Trains travel on the northbound track at an average rate of nine per hour and travel south at a rate of seven per hour.

a Find the probability that one or more trains travel on the northbound track and one or more trains travel on the southbound track in any randomly chosen 20-minute period.

b Find the probability that three or more trains pass the point on the train line in 20 minutes.

c Find the approximate probability that more than 150 trains pass the point in 10 hours.

11 X and Y are random variables such that $X \sim B(8, 0.4)$ and $Y \sim Po(4)$. Find:

a $E(4Y + 3X - 4)$

b $Var(2X - 4Y + 1)$

c $P(3X = 2Y + 18)$.

12 A machine makes a soft drink that is a mixture of juice concentrate and water.

For a medium cup, the amount of water has an average 125 ml and has a variance of 3. The amount of juice concentrate added is on average 10 ml with a variance of 6.

a Find the average volume and variance of a medium cup drink.

For a large cup of drink, 1.5 times more water and juice concentrate are used than for the medium cup.

b State the mean and variance of the volume of the large cup drink.

If a customer buys more than two but fewer than five cups of drink, a takeaway holder is used.

c Calculate the minimum average volume that a takeaway holder can take.

d Calculate the maximum average volume that a takeaway holder can take.

e Hence state the range of the average volume that a takeaway holder takes.

13 A biased die has the following probability distribution.

x	1	2	3	4	5	6
$P(X = x)$	k	$2k$	$3k$	$3k$	$2k$	k

a Find the expectation and variance of X.

Ellie rolls the die twice and the sum of the outcomes is recorded as a random variable Y.

b Find the expectation and variance of Y.

14 A local shop sells milk in three different sized bottles. One-pint bottles have a mean volume 568 ml and variance 12. Two-pint bottles have a mean volume 1137 ml and variance 26. Four-pint bottles have a mean volume 2268 ml and variance 38.

a Find the probability that two randomly chosen one-pint bottles have total volume less than a two-pint bottle.

b Find the probability that the combined volume of two one-pint bottles and one two-pint bottles exceeds that of a four-pint bottle. State one assumption made.

15 A shopkeeper claims that bags of sweets he sells each hold 150 g sweets. He knows the mean mass of sweets in a bag is 155 g and only 5% of bags contains less than 150 g.

  a Calculate the variance of the mass of sweets in a bag.

He sells gift boxes containing three bags of sweets. The mass of the gift box follows a normal distribution with mean 25 g and variance 8.

  b Find the probability that a gift box with three bags of sweets weighs more than 500 g.

  c Find the probability that the total weight of six bags of sweets, including the gift boxes, is between 950 g and 960 g.

16 A local market sells two different types of fish: seabass and salmon. On average, three packs of seabass and two packs of salmon are sold every two hours. Assume the sale of seabass and salmon follow a Poisson distribution and are independent of each other.

  a Find the mean and variance for the total packs of fish sold in 8 hours.

  b Find the probability that the market sold 25 packs of fish in 8 hours.

The manager of the local market is a fisherman. He goes fishing if more than 65 packs of fish are sold in a 24-hour period.

  c Find the probability that the manager goes fishing in a randomly chosen 24-hour period.

17 Two different types of flavoured microwave popcorn, sweet and salted, are sold in a shop. The probability that in a randomly chosen pack of sweet popcorn the corn doesn't pop completely is 0.8% and the corresponding probability for salted popcorn is 1.2%.

A party organiser bought 80 packs of each type of popcorn.

  a Using a suitable approximation, find the probability that exactly 158 packs of popcorn are completely popped.

  b The organiser buys equal numbers of both flavours. How many packs of each flavour must the party organiser buy to ensure that over 95% pop completely?

18 A music exam has two components: a practical test and a theory test. The marks for each test are normally distributed. The practical test has a mean of 75 and variance of 26. The theory test has a mean of 60 and variance of 12. The marks for both tests are then added together to give the overall grade.

  a Find the probability that a randomly chosen student scores more than 140.

The bottom 10% of students will not get a grade. The top 5% of students get an A* grade.

  b Find the lowest mark for which a student will get a grade.

  c Find the lowest mark for which a student will get an A* grade.

19 The random variables X and Y have the following distributions:

$X \sim \text{Po}(\gamma)$, $Y \sim \text{B}(50, 0.03)$ and it is known that $E(2X + 3Y) = 16.5$

  a Find the value of γ.

  b Find:

    i $E(2Y + X - 2)$

    ii $\text{Var}(3X - Y + 5)$

20 Two different taxi companies share the same taxi rank. On average, four taxis from Taxi Company A come to the rank every 10 minutes and three taxis from Taxi Company B come to the rank every 15 minutes.

a State an assumption that must be made in order for a Poisson distribution to be used.

b Find the probability that a total of 15 taxis come to the rank in a randomly chosen 20-minute period.

c Find the probability that from 16 to 20 taxis come to the rank in any 30-minute period.

d Find the probability that there is at most one more taxi from Taxi Company A than Taxi Company B in any 15-minute period.

Mathematics in life and work

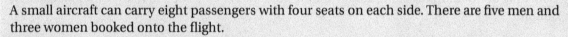

A small aircraft can carry eight passengers with four seats on each side. There are five men and three women booked onto the flight.

The mass, in kilograms, of men and women have distributions $N(79, 49)$ and $N(67, 28)$, respectively.

1 Find the probability that five men and three women weigh more than 650 kg in total.

For safety reasons the plane needs to be as balanced as possible. The seats are assigned so that there are four men on the left-hand side of the aisle and the rest of the passengers are on the right.

2 Find the probability that the combined mass of the men on the left of the aisle exceeds by more than 50 kg the combined mass of the passengers on the right.

3 CONTINUOUS RANDOM VARIABLES

Mathematics in life and work

In this chapter you will learn how to use continuous random variables to model situations. You will learn how to find the probability of a range of values as well as how to calculate the mean and standard deviation of a random variable.

The modelling of situations as continuous random variables can be used in many different careers, for example:

> If you were a manager of a train company you could model the times that trains are delayed as continuous random variables and find the expected delay time.

> If you were an advertiser you could calculate the median amount of time spent by a random person watching an online advert.

> If you were a manager of a mining company you could use continuous random variables to model the amount of ore that has been mined.

LEARNING OBJECTIVES

You will learn how to:

> use the properties of a probability density function

> use a probability density function to solve problems involving probabilities

> use a probability density function to calculate the mean and variance of a distribution.

LANGUAGE OF MATHEMATICS

Key words and phrases you will meet in this chapter:

> continuous random variable, piecewise function, probability density function

PREREQUISITE KNOWLEDGE

You should already know how to:

> calculate the expectation and variance of a discrete random variable

> integrate a function.

You should be able to complete the following questions correctly:

1 X is a random variable with probability distribution function:

x	0	1	2	3	4
$P(X = x)$	0.2	0.3	0.25	p	0.15

   a Find p.   **b** Find $E(X)$ and $\text{Var}(X)$.

2 **a** Calculate $\int_1^4 (2x + 3)\,dx$.

   b Find the positive value of a such that $\int_0^a \left(\frac{x}{2} + \frac{1}{2}\right)dx = 2$.

3.1 Probability density functions

In **Probability and Statistics 1**, **Chapter 4**, **The normal distribution** you learnt about the normal distribution, which is an example of a **continuous random variable**. The normal distribution is the most common distribution used to describe continuous random variables, but it is not the only one.

If you were to arrive at a random time at a railway station and then plot the length of time you wait until the first train arrives, you could get a graph like those below.

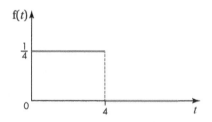 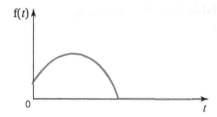

Stop and think   What does the first graph suggest about the arrival times of the trains?

As with a normal distribution, the area under each graph shows the probability. This means that the total area under the graphs is 1. You can calculate the probability that you would be waiting less than 2 minutes by calculating the area under the graph between 0 and 2. For the first graph this would be easy to calculate:

$$P(X < 2) = 2 \times \frac{1}{4} = 0.5$$

For the second graph you would use integration to find the area under the graph. To do this, you would need to know the equation of the **probability density function**.

KEY INFORMATION

All continuous random variables have probability distribution functions.

Continuous random variables often have their probability density function defined by a **piecewise function**. The second graph has the following probability density function:

$$f(x) = \begin{cases} \dfrac{3}{56}(4 + 3x - x^2)\,dx & 0 \leqslant x \leqslant 4, \\ 0 & \text{otherwise.} \end{cases}$$

Since $P(X \leqslant 2)$ is represented by the area under the curve, you can integrate using the limits $x = 0$ and $x = 2$.

$$P(X \leqslant 2) = \int_0^2 \frac{3}{56}\left(4 + 3x - x^2\right) dx$$

$$= \frac{3}{56}\int_0^2 \left(4 + 3x - x^2\right) dx$$

$$= \frac{3}{56}\left[4x + \frac{3x^2}{2} - \frac{x^3}{3}\right]_0^2$$

$$= \frac{3}{56}\left[\left(4(2) + \frac{3(2)^2}{2} - \frac{(2)^3}{3}\right) - \left(4(0) + \frac{3(0)^2}{2} - \frac{(0)^3}{3}\right)\right]$$

$$= \frac{3}{56}\left[\frac{34}{3}\right]$$

$$= 0.607$$

For any value of x outside the interval [0, 4], the value of the function is 0.

KEY INFORMATION

$$P(a < X < b) = \int_a^b f(x)\,dx$$

Example 1

The continuous random variable X has the following probability density function:

$$f(x) = \begin{cases} k(x + 1) & 1 \leqslant x \leqslant 3, \\ 0 & \text{otherwise.} \end{cases}$$

where k is a constant.

a Find k.

b Sketch the probability distribution function.

c Find $P(X < 1.5)$.

Solution

a $\int_1^3 k(x + 1)\,dx = 1$

$$k\int_1^3 (x + 1)\,dx = 1$$

$$k\left[\frac{x^2}{2} + x\right]_1^3 = 1$$

$$k\left[\left(\frac{(3)^2}{2} + 3\right) - \left(\frac{(1)^2}{2} + 1\right)\right] = 1$$

Remember that the area under the curve represents the probability, so it must sum to 1.

$6k = 1$

$k = \dfrac{1}{6}$

Therefore $f(x) = \begin{cases} \dfrac{1}{6}(x+1) & 1 < x < 3, \\ 0 & \text{otherwise.} \end{cases}$

b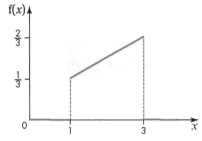

c $P(X < 1.5) = \displaystyle\int_1^{1.5} \dfrac{1}{6}(x+1)\,dx$

$= \dfrac{1}{6}\displaystyle\int_1^{1.5}(x+1)\,dx$

$= \dfrac{1}{6}\left[\dfrac{x^2}{2} + x\right]_1^{1.5}$

$= \dfrac{1}{6}\left[\left(\dfrac{(1.5)^2}{2} + 1.5\right) - \left(\dfrac{(1)^2}{2} + 1\right)\right]$

$= \dfrac{3}{16}$

The shaded area on the probability density graph below also represents $P(X < 1.5)$

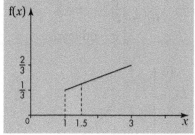

Example 2

A continuous random variable L has the following probability density function:

$f(l) = \begin{cases} \dfrac{k}{l^3} & l \geqslant 2, \\ 0 & \text{otherwise.} \end{cases}$

where k is a constant.

a Find k.

b Calculate $P(L < 5)$.

Solution

a $\int_2^\infty \frac{k}{l^3} dl = 1$

$k \int_2^\infty \frac{1}{l^3} dl = 1$

$k \left[\frac{-1}{2l^2} \right]_2^\infty = 1$

$-\frac{k}{2} \left[\frac{1}{l^2} \right]_2^\infty = 1$

$-\frac{k}{2} \left(0 - \frac{1}{4} \right) = 1$

$\frac{k}{8} = 1$

$k = 8$

Therefore:

$$f(l) = \begin{cases} \dfrac{8}{l^3} & l \geqslant 2, \\ 0 & \text{otherwise.} \end{cases}$$

> There is no upper bound for f(l) so the upper limit must be infinity.

> As $l \to \infty \; \frac{1}{l^2} \to 0$

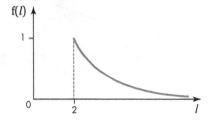

b $P(L < 5) = \int_2^5 \frac{8}{l^3} dl$

$= \left[\frac{8}{-2l^2} \right]_2^5$

$= \left[\frac{-4}{l^2} \right]_2^5$

$= \left[\frac{-4}{5^2} - -\frac{4}{2^2} \right]$

$= -\frac{4}{25} + 1$

$= \frac{21}{25}$

$= 0.84$

Exercise 3.1A

C **1** For each of these functions, state with a reason whether or not they could be a probability density function of a continuous random variable X.

a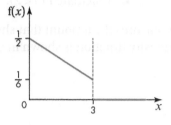

b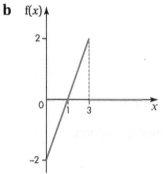

c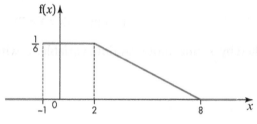

2 T is a continuous random variable with the following probability density function:

$$f(t) = \begin{cases} k(t-1) & 1 \leqslant t \leqslant 5, \\ 0 & \text{otherwise.} \end{cases}$$

a Find the value of k.　　　　**b** Find $P(T < 3)$.　　　　**c** Find $P(2 < T \leqslant 4)$.

3 The continuous random variable X has the following probability density function:

$$f(x) = \begin{cases} kx^2 & -2 \leqslant x \leqslant 2, \\ 0 & \text{otherwise.} \end{cases}$$

a Find the value of k.　　　　　　　**b** Sketch the graph of the probability density function.

c Write down $P(X = 1)$.　　　　　　**d** Find $P(X \leqslant 1)$.

4 The continuous random variable S has the following probability density function:

$$f(s) = \begin{cases} ks(3-s) & 0 \le s \le 3, \\ 0 & \text{otherwise.} \end{cases}$$

a Sketch the graph of f(s).   **b** Find the value of k.   **c** Calculate $P(S < 2)$.

5 A scientist uses a pipette to measure out 20 ml of liquid. She measures the amount that she actually uses and records the error, X, in ml. The probability density function is shown in the graph below.

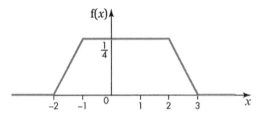

a Find the probability that the error is between −1 and 1.

b Find the probability that she measures out more than 20 ml of liquid.

6 A continuous random variable, X, has the following probability density function:

$$f(x) = \begin{cases} \dfrac{x}{8} & 0 \le x \le a, \\ 0 & \text{otherwise.} \end{cases}$$

a Calculate the value of a.   **b** Find $P(X > 2)$.   **c** Find $P(3 < X < 7)$.

7 The delay of a train, in minutes, can be modelled by a continuous random variable, T, with the following probability density function:

$$f(t) = \begin{cases} \dfrac{k}{\sqrt{t^3}} & t \ge 1, \\ 0 & \text{otherwise.} \end{cases}$$

a Find the value of k.

b Calculate $P(T < 4)$.

c Calculate the probability that the delay is longer than 9 minutes, given that it is already delayed by 4 minutes.

8 A set of minor roadworks delays drivers' journeys. The time, T minutes, by which their journeys are delayed is a continuous random variable with a probability density function defined by

$$f(t) = \begin{cases} 5kt & 0 \le t \le 20, \\ 0 & \text{otherwise.} \end{cases}$$

where k is a constant.

a Find the value of k.

b Find the probability that a person, selected at random, will experience a delay of:

   i exactly 10 minutes

   ii between 5 and 10 minutes.

9 The continuous random variable X has the following probability density function:

$$f(x) = \begin{cases} kx & 0 \leqslant x \leqslant a, \\ 0 & \text{otherwise.} \end{cases}$$

Given that $P(X < 2) = \frac{2}{9}$, calculate:

a the value of k and a **b** $P(1 < X < 4)$.

10 A continuous random variable X has the following probability density function:

$$f(x) = \begin{cases} kx^2 & -1 \leqslant x \leqslant 1, \\ k(2 - x) & 1 < x \leqslant 2, \\ 0 & \text{otherwise.} \end{cases}$$

a Sketch a graph of this function.   **b** Find the value of k.

c Calculate $P(0.5 < X < 2)$.

Mathematics in life and work: Group discussion

If you were a teacher setting a maths examination, you might need to know how long it takes to answer a particular set of mathematics problems. The continuous random variable T (the time taken) could be modelled using a probability density function.

1 What shape function would be appropriate for this model? What shape functions would be inappropriate? Justify your answers.

2 It is decided that one of the questions in the examination is too difficult and needs to be removed. Is it likely that this would affect the shape of the original function?

3 Once a model has been decided upon, how might it be used in practice to set the length of time for a maths examination?

3.2 Median and percentiles of a continuous random variable

In **Probability and Statistics 1, Chapter 1, Representation of data** you saw that the median is the middle value of a set of data, where half of the data is above the median and half is below. This can be extended to continuous random variables.

Consider the continuous random variable X with probability density function $f(x)$. The median is the value of X that divides the area under the curve in half.

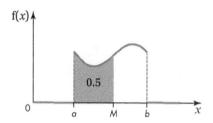

This can be calculated as $P(X \leqslant M) = \int_a^M f(x)\mathrm{d}x = \frac{1}{2}$, where M is the median.

Using a similar approach, you can calculate percentiles of the distribution of X.

The 70th percentile is the value, x, such that $P(X \leqslant x) = 70\%$.

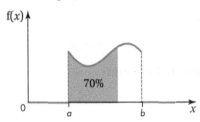

KEY INFORMATION

The Nth percentile of a continuous random variable is the value n such that

$$P(X \leqslant n) = \int_a^n f(x)\,dx = \frac{N}{100}$$

Example 3

The continuous random variable X has the following probability density function:

$$f(x) = \begin{cases} \frac{1}{50}(10 - x) & 0 \leqslant x \leqslant 10, \\ 0 & \text{otherwise.} \end{cases}$$

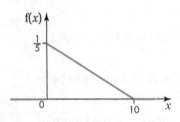

a Find the median.

b Find the upper quartile.

Solution

a Using the integration method:

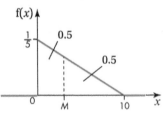

The median is a value of x so that the area under the curve is 0.5.

$$P(X < M) = \int_0^M \frac{1}{50}(10 - x)\,dx = \frac{1}{2}$$

$$\int_0^M \left(\frac{1}{5} - \frac{x}{50}\right)dx = \frac{1}{2}$$

$$\left[\frac{x}{5} - \frac{x^2}{100}\right]_0^M = \frac{1}{2}$$

$$\left(\frac{M}{5} - \frac{M^2}{100}\right) - (0) = \frac{1}{2}$$

$$20M - M^2 = 50$$

$$M^2 - 20M + 50 = 0$$

Using the quadratic formula:

$$M = \frac{20 \pm \sqrt{(-20)^2 - 4(1)(50)}}{2}$$

$$M = 10 \pm 5\sqrt{2}$$

$$M = 2.929\ldots \text{ or } 17.071\ldots$$

Since M is between 0 and 10, the median is 2.93 (3 s.f.).

Using the geometric method:

When $x = M$, $y = \frac{1}{50}(10 - M)$

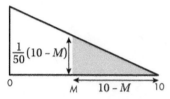

Area of shaded triangle $= \frac{1}{2} \times (10 - M) \times \frac{1}{50}(10 - M)$

$$= \frac{1}{100}(10 - M)^2$$

Since $P(X > M) = 0.5$ then the area of the triangle $= 0.5$

So $\frac{1}{100}(10 - M)^2 = 0.5$

$$(10 - M)^2 = 50$$

Taking the square root on both sides:

$$10 - M = \pm\sqrt{50}$$

$$M = 10 \pm \sqrt{50}$$

$$M = 2.929 \text{ or } 17.071$$

Reject the second value, because $M = 17.071$ will result in a negative value for $10 - M$.

Therefore the median is 2.93 (3 s.f.).

b The upper quartile is the same as the 75th percentile.

$$P(X < U) = \int_0^U \frac{1}{50}(10 - x)\,dx = \frac{75}{100}$$

$$\int_0^U \left(\frac{1}{5} - \frac{x}{50}\right)dx = \frac{75}{100}$$

$$\left[\frac{x}{5} - \frac{x^2}{100}\right]_0^U = \frac{75}{100}$$

$$\left(\frac{U}{5} - \frac{U^2}{100}\right) - (0) = \frac{75}{100}$$

$$20U - U^2 = 75$$

$$U^2 - 20U + 75 = 0$$

Using the quadratic formula:

$$U = \frac{20 \pm \sqrt{(-20)^2 - 4(1)(75)}}{2}$$

$$U = 15 \text{ or } 5$$

Since U is between 0 and 10, the upper quartile is 5.

Using the geometric method, when $x = U$, $y = \frac{1}{50}(10 - U)$.

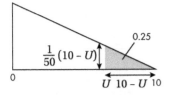

Since $P(X \leq U) = 0.75$ then the area of the triangle is $1 - 0.75 = 0.25$

$$\text{Area of shaded triangle} = \frac{1}{2} \times (10 - U) \times \frac{1}{50}(10 - U)$$

$$= \frac{1}{100}(10 - U)^2$$

So $\frac{1}{100}(10 - U)^2 = 0.25$

$$(10 - U)^2 = 25$$

Taking the square root on both sides:

$$10 - U = \pm 5$$

$$U = 5 \text{ or } 15$$

Reject $U = 15$ because this will result in a negative value for $10 - U$.

The upper quartile is 5.

Stop and think   Is it always true that dividing a right-angled triangle into two parts, using the perpendicular bisector of the base, results in the area being spilt in the ratio 1:3?

Exercise 3.2A

1 The continuous random variable X has a probability density function given by:

$$f(x) = \begin{cases} \dfrac{1+2x}{k} & 0 \leqslant x \leqslant 4, \\ 0 & \text{otherwise.} \end{cases}$$

a Find the value of k. **b** Calculate the median.

c Calculate the 30th percentile.

2 The continuous random variable T has a probability density function given by:

$$f(t) = \begin{cases} \dfrac{1}{64}t^3 & 0 \leqslant t \leqslant 4, \\ 0 & \text{otherwise.} \end{cases}$$

a Find the median. **b** Calculate the 90th percentile.

(C) 3 The continuous random variable W has a probability density function given by:

$$f(w) = \begin{cases} \dfrac{3}{14}\left(w^2+2\right) & -1 \leqslant w \leqslant 1, \\ 0 & \text{otherwise.} \end{cases}$$

a Without doing any calculations, state the median. Explain your answer.

b State, with a reason, whether the upper quartile of this distribution is smaller than or greater than 0.6.

(MM) 4 The length of time, in 100s of hours, that a light bulb lasts can be modelled as a continuous random variable with the following probability density function:

$$f(h) = \begin{cases} \dfrac{18}{h^3} & h \geqslant 3, \\ 0 & \text{otherwise.} \end{cases}$$

a Calculate, in hours, the median length of time the bulb lasts.

The company wants to create a guarantee policy which will replace 4% of bulbs.

b Calculate, in hours, the longest lifespan of a bulb that will be replaced under this policy.

5 The continuous random variable X has a probability density function given by:

$$f(x) = \begin{cases} \dfrac{x^3}{k}\left(16 - x^4\right) & 0 \leqslant x \leqslant 2, \\ 0 & \text{otherwise.} \end{cases}$$

a Find the value of k. **b** Find the median.

(PS) 6 The continuous random variable Y has a probability density function given by:

$$f(y) = \begin{cases} \dfrac{k}{y} & 1 \leqslant y \leqslant a, \\ 0 & \text{otherwise.} \end{cases}$$

The median of Y is $\sqrt{5}$. Find the values of k and a.

7 The continuous random variable Y has a probability density function given by:

$$f(y) = \begin{cases} 2(1-y) & k \leqslant y \leqslant 1, \\ 0 & \text{otherwise.} \end{cases}$$

a Find the value of k.

b Sketch the graph of the probability density function.

c Find the 90th percentile.

8 A continuous random variable has the following probability density function graph:

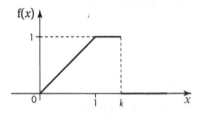

a Find the value of k.

b Find the median.

c Find the 80th percentile.

9 The length of time, T, in hours, that a candle burns can be modelled as a continuous random variable with the following probability density function:

$$f(t) = \begin{cases} \frac{1}{5}\left(1.75 - t^2\right) + \frac{34}{15} & 1.5 \leqslant t \leqslant 2, \\ 0 & \text{otherwise.} \end{cases}$$

a Calculate the median length of time, in hours, the candle burns.

b A restaurant manager purchased one hundred candles for an evening function. How many candles will burn for less than 100 minutes?

10 The continuous random variable X has a probability density function given by:

$$f(x) = \begin{cases} \frac{1}{6}x - 1 & 6 \leqslant x \leqslant a, \\ 0 & \text{otherwise.} \end{cases}$$

a Find the value of a.

b Find the median.

3.3 Mean and variance of a continuous random variable

Mean of a continuous random variable

In **Probability and Statistics 1, Chapter 3, Discrete random variables** you learnt that the mean (expectation) of a discrete random variable is given by $E(X) = \Sigma xp$. The equivalent formula for the mean of a continuous random variable is $E(X) = \int_{-\infty}^{\infty} x\,f(x)\,dx$.

> The mean (expectation) can also be denoted by μ.

Example 4

The random variable X has a probability density function given by:

$$f(x) = \begin{cases} \dfrac{1}{18}(3+x) & -3 \leqslant x \leqslant 3, \\ 0 & \text{otherwise.} \end{cases}$$

Find the mean of X.

Solution

$$E(X) = \int_{-3}^{3} x \times \frac{1}{18}(3+x)\,dx$$

$$= \frac{1}{18}\int_{-3}^{3}(3x + x^2)\,dx$$

$$= \frac{1}{18}\left[\frac{3x^2}{2} + \frac{x^3}{3}\right]_{-3}^{3}$$

$$= \frac{1}{18}\left[\left(\frac{3 \times 3^2}{2} + \frac{3^3}{3}\right) - \left(\frac{3 \times (-3)^2}{2} + \frac{(-3)^3}{3}\right)\right]$$

$$= \frac{1}{18}\left[\frac{45}{2} - \frac{9}{2}\right]$$

$$= 1$$

The mean of X is 1.

> **KEY INFORMATION**
>
> The mean of X is given by
> $$\int_{-\infty}^{\infty} x\,f(x)\,dx$$

> The probability density function only takes non-zero values between −3 and 3, so you can use −3 and 3 as limits.

Stop and think If the probability density function is symmetrical, what can you say about the mean?

Variance of a continuous random variable

As with the mean, you can adapt the variance formula from **Probability and Statistics 1, Chapter 3, Discrete random variables** for use with continuous random variables.

$Var(X) = E(X^2) - \{E(X)\}^2$ leads to the formula $Var(X) = \int_{-\infty}^{\infty} x^2 f(x)\,dx - \mu^2$ where $\mu = E(X)$.

> **KEY INFORMATION**
>
> For a continuous random variable,
> $$Var(X) = \int_{-\infty}^{\infty} x^2 f(x)\,dx - \mu^2,$$
> where $\mu = E(X)$.

Example 5

The continuous random variable X has the following probability density function:

$$f(x) = \begin{cases} 1 - \dfrac{x}{2} & 0 \leqslant x \leqslant 2, \\ 0 & \text{otherwise.} \end{cases}$$

a Find the mean and standard deviation of X.

b Find $P(X > \mu + \sigma)$.

Solution

a $E(X) = \int_0^2 x\left(1 - \frac{x}{2}\right)dx$

$= \int_0^2 \left(x - \frac{x^2}{2}\right)dx$

$= \left[\frac{x^2}{2} - \frac{x^3}{6}\right]_0^2$

$= \left(\frac{2^2}{2} - \frac{2^3}{6}\right) - (0)$

$= \frac{2}{3}$

$Var(X) = \int_0^2 x^2\left(1 - \frac{x}{2}\right)dx - \left(\frac{2}{3}\right)^2$

$= \int_0^2 \left(x^2 - \frac{x^3}{2}\right)dx - \left(\frac{2}{3}\right)^2$

$= \left[\frac{x^3}{3} - \frac{x^4}{8}\right]_0^2 - \left(\frac{2}{3}\right)^2$

$= \left(\frac{2^3}{3} - \frac{2^4}{8}\right) - (0) - \left(\frac{2}{3}\right)^2$

$= \frac{2}{9}$

Therefore the standard deviation is $\sigma = \sqrt{\frac{2}{9}} = 0.471$ (3 s.f.).

b $P(X > \mu + \sigma) = P\left(X > \frac{2}{3} + 0.4714\right)$

$= P(X > 1.1381)$

$= \int_{1.1381}^2 \left(1 - \frac{x}{2}\right)dx$

$= \left[x - \frac{x^2}{4}\right]_{1.1381}^2$

$= \left(2 - \frac{2^2}{4}\right) - \left(1.1381 - \frac{(1.1381)^2}{4}\right)$

$= 0.186$

> **KEY INFORMATION**
>
> The variance of a continuous random variable is given by
>
> $$Var(X) = \int_{-\infty}^{\infty} x^2 f(x)\,dx - \{E(X)\}^2$$

Exercise 3.3A

1 The continuous random variable X has a probability density function given by:

$$f(x) = \begin{cases} \dfrac{x^3}{320} & 2 \leqslant x \leqslant 6, \\ 0 & \text{otherwise.} \end{cases}$$

a Find the expectation.   **b** Find the variance.

2 The continuous random variable X has a probability density function given by:

$$f(x) = \begin{cases} 2x - 4 & 2 \leqslant x \leqslant 3, \\ 0 & \text{otherwise.} \end{cases}$$

a Find the mean.   **b** Find the standard deviation.

3 The continuous random variable X has a probability density function given by:

$$f(x) = \begin{cases} k\sqrt{x} & 4 \leqslant x \leqslant 9, \\ 0 & \text{otherwise.} \end{cases}$$

a Find k.   **b** Find the expectation and variance of X.

MM 4 The mass, W, in grams, of leftover food on a plate at the end of a meal is modelled by a continuous random variable with the following probability density function:

$$f(w) = \begin{cases} \dfrac{k}{\sqrt{w^7}} & w \geqslant 1, \\ 0 & \text{otherwise.} \end{cases}$$

a Find the value of k.

b What is the expected amount of food left on a plate?

c Find the standard deviation of W.

5 The continuous random variable X has a probability density function given by:

$$f(x) = \begin{cases} \dfrac{1}{2\sqrt{x}} & 1 \leqslant x \leqslant 4, \\ 0 & \text{otherwise.} \end{cases}$$

a Find the mean.   **b** Find the variance.   **c** Find $P(X < \mu)$.

PS 6 The continuous random variable X has a probability density function given by:

$$f(x) = \begin{cases} \dfrac{1}{c} & 0 \leqslant x \leqslant c, \\ 0 & \text{otherwise.} \end{cases}$$

a Find the mean in terms of c.

b Find the variance in terms of c.

c Find $P(X > \mu)$ if $c = 4$.

7 The continuous random variable X has a probability density function given by:

$$f(x) = \begin{cases} k(2x+4) & -1 \leqslant x \leqslant 3, \\ 0 & \text{otherwise.} \end{cases}$$

a Find the value of k.

b Find the standard deviation.

c Find the probability that a randomly chosen value from this distribution lies within one standard deviation of the mean.

8 Megan goes to the gym every day. The exercising time, T, in hours, is modelled by a continuous random variable with the following probability density function:

$$f(t) = \begin{cases} 1 - \frac{3}{8}t^2 & 0 \leqslant t \leqslant 2, \\ 0 & \text{otherwise.} \end{cases}$$

a How many hours does Megan exercise every day on average?

b Calculate the variance and then comment on if Megan is consistent with her exercise.

9 The waiting time, T, in minutes, for a customer to be served at a counter is modelled by a continuous random variable with the following probability density function:

$$f(t) = \begin{cases} kt(6-t) & 0 \leqslant t \leqslant 6, \\ 0 & \text{otherwise.} \end{cases}$$

a Find the value of k.

b Sketch the graph of $f(t)$.

c What is the mean waiting time for a customer to be served?

10 The continuous random variable X has the following probability density function:

$$f(x) = \begin{cases} \frac{8}{9}x^2 & 0 \leqslant x \leqslant p, \\ 0 & \text{otherwise.} \end{cases}$$

a Find the value of p.

b Calculate $P(X < \text{Mean})$.

Mathematics in life and work: Group discussion

You are working for an ore-mining company. It is known that each load will contain other materials such as rock as well as the ore that the company sells. The amount of unusable materials, in kg, per load is given by X and can be modelled as a continuous random variable with the following probability density function:

$$f(x) = \begin{cases} \dfrac{3}{32000}(x-10)(50-x) & 10 \leqslant x \leqslant 50, \\ 0 & \text{otherwise.} \end{cases}$$

1 Find the mean and standard deviation of the amount of unusable materials.

The company sell a load of ore that has been mined for \$850, but they will only get \$580 per load if it contains more than 37.5 kg of unusable material.

2 Find the expected value of a full load of mined ore.

The company could buy a machine that sorts the mined materials. This would mean there was a very low amount of unusable material so the load could be sold for \$940. You know that you will only be able to mine 75% of the original output by using the company resources for buying and manning the machine.

3 Should the company use the machine to sort the loads? If so, should the machine sort all or some of the loads to sell?

SUMMARY OF KEY POINTS

> A continuous random variable has a probability density function such that $P(X \leqslant x) = \int_{-\infty}^{x} f(x) dx$.

> The median, M, of a continuous random variable is given by $\int_{-\infty}^{M} f(x) dx = 0.5$.

> The pth percentile is given by $\int_{-\infty}^{P} f(x) dx = \dfrac{p}{100}$.

> The mean is given by $E(X) = \int_{-\infty}^{\infty} x f(x) dx$.

> The variance is given by $\text{Var}(X) = E(X^2) - (E(X))^2 = \int_{-\infty}^{\infty} x^2 f(x) dx - \{E(X)\}^2$.

EXAM-STYLE QUESTIONS

1 The continuous random variable, X, has a probability density distribution as shown below. It is given that $P(X < c) = 0.68$.

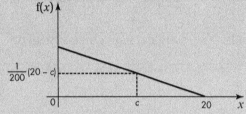

a Find c.

b Calculate $P(10 < X < 12)$.

2 The time, T, in minutes, between phone calls being made to a company can be modelled by a continuous random variable with the following probability density function:

$$f(t) = \begin{cases} \dfrac{1}{k} & 0 \leqslant t \leqslant 10, \\ 0 & \text{otherwise.} \end{cases}$$

a Find the value of k.

b State the median.

c Find the probability that the time between phone calls is between 3 and 8 minutes.

3 A random variable L represents the amount of liquid in a container and has the following probability density function:

$$f(l) = \begin{cases} kl & 0 \leqslant l \leqslant 10, \\ 0 & \text{otherwise.} \end{cases}$$

a Find k.

b Find: **i** $P(L < 6)$ **ii** $P(3 < L < 8)$.

4 The waiting time, in minutes, for a bus is modelled by a continuous random variable, W. The probability density function is given by:

$$f(w) = \begin{cases} k(w+1) & 0 \leqslant w \leqslant 10, \\ 0 & \text{otherwise.} \end{cases}$$

a Show that $k = \frac{1}{60}$.

b Find the probability that the waiting time is less than 3 minutes.

c Find the median value for X.

5 A continuous random variable, X, has a probability density function given by:

$$f(x) = \begin{cases} kx^2 & 0 \leqslant x \leqslant 3, \\ 0 & \text{otherwise.} \end{cases}$$

a Show that $k = \frac{1}{9}$. **b** Sketch the graph of f(x). **c** Find $P(X > 1)$.

6 The continuous random variable, X, has a probability density function given by:

$$f(x) = \begin{cases} kx & 0 \leqslant x \leqslant 4, \\ 0 & \text{otherwise.} \end{cases}$$

a Find the value of k. **b** Find the median of X. **c** Find the mean of X.

7 The average speed, in ms^{-1}, of a randomly selected car on a clear stretch of road can be modelled as a continuous random variable with the following probability density function:

$$f(x) = \begin{cases} k & 28 \leqslant x \leqslant 34, \\ 0 & \text{otherwise.} \end{cases}$$

a Find k.

b Sketch the graph of f(x) and hence state the mean of X.

c Using integration, calculate the variance of X.

8 The time spent, in minutes, waiting in a queue can be modelled by a random variable, Q, with the following probability density function:

$$f(q) = \begin{cases} kq + 0.06 & 1 \leqslant q \leqslant 6, \\ 0 & \text{otherwise} \end{cases}$$

where k is a positive constant.

a Sketch the graph of f(q).

b Show that the value of k is $\frac{1}{25}$.

c Find the probability the wait time is between 3 and 5 minutes.

d Calculate the interquartile range of the wait time.

C **9** The lifetime, t, in months, of an electrical component can be modelled by a continuous
random variable with the following probability density function:

$$f(t) = \begin{cases} \dfrac{k}{t^2} & 20 \leqslant t \leqslant 28, \\ 0 & \text{otherwise.} \end{cases}$$

a Find k.

b Find the probability a randomly chosen component has a lifetime of over 2 years.

c Find the expected value of T.

d Find $P(T \leqslant E(T))$.

e Hence explain whether the median is smaller than or greater than the mean
of this distribution.

C **10** The time taken for a randomly chosen student to answer a question on an online test can be
modelled by a continuous random variable X with the following probability density function:

$$f(x) = \begin{cases} k\sqrt{x} & 1 \leqslant x \leqslant 4, \\ 0 & \text{otherwise.} \end{cases}$$

a Suggest what could happen at $x = 4$.

b Find k.

c Calculate: **i** the median **ii** the mean.

d Find the probability that a student takes between the mean and median time
to answer the question.

MM **11** A continuous random variable has the following probability density function:

$$f(x) = \begin{cases} x(3a - x) & 0 \leqslant x \leqslant 2a, \\ 0 & \text{otherwise.} \end{cases}$$

a Show that $a^3 = \dfrac{3}{10}$. **b** Find $P(X < 0.5)$. **c** Find $E(X)$.

12 The time, in hours, spent playing video games in a day can be modelled by a continuous
random variable, X, which has the following probability density function:

$$f(x) = \begin{cases} a e^{-ax} & x \geqslant 0, \\ 0 & \text{otherwise} \end{cases}$$

where a is a positive constant.

a Verify that this is a properly defined probability density function.

b Find the median time spent playing video games in terms of a.

It is found that $a = 0.3$.

c Calculate the probability that a gamer spends more than 5 hours playing in one day.

13 The random variable X has a probability density function defined by:

$$f(x) = \begin{cases} \dfrac{1}{18}(x-4)^2 & a \leqslant x \leqslant 7. \\ 0 & \text{otherwise.} \end{cases}$$

a Show that $a = 1$.

b Find the values for the median and the lower quartile of X.

c Determine $P(2 \leqslant x \leqslant 3)$.

14 The continuous random variable X has a probability density function defined by:

$$f(x) = \begin{cases} k & -3k \leqslant x \leqslant 6k \\ 0 & \text{otherwise,} \end{cases}$$

a i Sketch the graph of $f(x)$.

  ii Hence show that $k = \dfrac{1}{3}$.

b Find the exact numerical value for the mean and the standard deviation of X.

c Find $P(X \geqslant 1)$.

15 The continuous random variable Y has probability density function given by:

$$f(y) = \begin{cases} \dfrac{1}{2}\left(y^2 + 2\right) - \dfrac{1}{6} & 0 \leqslant y \leqslant 1, \\ 0 & \text{otherwise.} \end{cases}$$

a Sketch the graph of $f(y)$.

b Calculate $P(Y \leqslant 0.6)$.

c Show that $E(Y^2) = \dfrac{17}{45}$

d Given that $E(Y) = \dfrac{13}{24}$, find $\text{Var}(Y)$.

16 The error, in minutes, made by Tom in estimating the time that he takes to complete an assignment may be modelled by the random variable T, with the probability density function:

$$f(t) = \begin{cases} \dfrac{1}{20} & -5 \leqslant t \leqslant k, \\ 0 & \text{otherwise.} \end{cases}$$

a Show that $k = 15$.

b Find: **i** $E(T)$ **ii** $\text{Var}(T)$.

c Calculate the probability that Tom will make an error of magnitude of at least 3 minutes when estimating the time that he takes to complete a given assignment.

17 The number of hours, X, needed to service a machine can be modelled by the following probability density function:

$$f(x) = \begin{cases} \dfrac{-2}{9}(x-1)(x-4) & 1 \leqslant x \leqslant 4, \\ 0 & \text{otherwise.} \end{cases}$$

a Find the mean of X.

b Find the probability that X exceeds 2.5 hours.

An additional charge will be applied if the service takes more than 3.5 hours.

c Find the probability that an additional charge is applied.

18 The queuing time, X minutes, of vehicles at a petrol station has probability density function:

$$f(x) = \begin{cases} \dfrac{6}{125} x(k-x) & 0 \leqslant x \leqslant k, \\ 0 & \text{otherwise.} \end{cases}$$

a Show that the value of k is 5.

b Write down the value of $E(X)$.

c Calculate $\text{Var}(X)$.

d Find the probability that the queuing time of a randomly chosen vehicle will differ from the mean by at least half a minute.

19 A continuous random variable X has the probability density function $f(x)$ shown below.

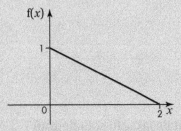

a Show that $f(x) = 1 - \dfrac{1}{2}x$ for $0 \leqslant x \leqslant 2$ and specify $f(x)$ for all real values of X.

b Write down the value of $P(X = 1)$.

c Find the median of X.

d Show that the 80th percentile of X lies between 1.105 and 1.115.

20 The lifetime, H, in tens of hours, of a rechargeable battery has a probability density function f(h) given by:

$$f(h) = \begin{cases} k(3h^2 + 2h - 5) & 1 \leqslant h \leqslant 3, \\ 0 & \text{otherwise.} \end{cases}$$

  a Show that $k = \dfrac{1}{24}$.

  b Find the median of H, giving your answer to 3 significant figures.

  c Find $P(H \geqslant 2.4)$.

A string of lights is powered by two batteries, both of which must be working. Two new batteries are put into the string of lights.

  d Find the probability that the string of lights will still be working after 24 hours.

21 The random variable X has probability density function f(x), where a is a constant.

$$f(x) = \begin{cases} \dfrac{a}{1 + x^2} & 0 \leqslant x \leqslant 1, \\ 0 & \text{otherwise.} \end{cases}$$

  a Show that $a = \dfrac{4}{\pi}$.

  b Show that $E(X) = \dfrac{b \ln c}{\pi}$, where $b, c \in \mathbb{N}$.

22 A probability density function is given by:

$$f(x) = \begin{cases} k\mathrm{e}^x \sin x & 0 \leqslant x \leqslant \dfrac{\pi}{2}, \\ 0 & \text{otherwise.} \end{cases}$$

  a Find the value of k.

  b Find $P\left(0 < x < \dfrac{\pi}{4}\right)$.

Mathematics in life and work

You are the manager of a mine. A piece of machinery that is used to blast out the rock runs perfectly for the first year but it is known that it will fail within the first 4 years. The length of time, in years, that the machine works for is modelled by a continuous random variable, X, which has the following probability density function:

$$f(x) = \begin{cases} \dfrac{k}{x} & 1 \leqslant x \leqslant 4 \\ 0 & \text{otherwise.} \end{cases}$$

1 Show that $k = \dfrac{1}{2 \ln 2}$.

2 Find the median lifetime of the machine.

3 Find the probability that the machine lasts less than 3 years.

If the machine fails after the first year you will have to repair the machine for $2500. The company that sells the machines offers a guarantee that covers the cost of repairing the machine if it fails within the first 3 years. The cost of this guarantee is $1800.

4 Should you buy the guarantee? Explain your answer.

4 SAMPLING AND ESTIMATION

Mathematics in life and work

It is often useful to collect data from a sample of a population and use this to make inferences about the whole population. Data sampling is used in a range of careers, for example:

- If you were a quality control supervisor at a light bulb manufacturer, you could use a sample of bulbs to find out how long the light bulbs are expected to last.

- If you were a fisheries manager, you could count the fish in a certain area to estimate the size of the whole population.

- If you were a political campaign manager, you could use a sample from the voting population to find out whether your tactics were increasing support for your candidate.

LEARNING OBJECTIVES

You will learn how to:

- understand the distinction between a sample and a population, and appreciate the necessity for randomness in choosing samples
- explain in simple terms why a given sampling method may be unsatisfactory
- recognise that a sample mean can be regarded as a random variable, and use the facts that

  $E(\bar{X}) = \mu$ and $Var(\bar{X}) = \dfrac{\sigma^2}{n}$

- use the fact that $\bar{X}$ has a normal distribution if X has a normal distribution
- use the Central Limit Theorem where appropriate
- calculate unbiased estimates of the population mean and variance from a sample, using either raw or summarised data
- determine and interpret a confidence interval for a population mean in cases where the population is normally distributed with known variance or where a large sample is used
- determine, from a large sample, an approximate confidence interval for a population proportion.

LANGUAGE OF MATHEMATICS

Key words and phrases you will meet in this chapter:

- bias, unbiased confidence interval, confidence level, point estimate, population parameter, population proportion, sampling distribution, standard deviation, standard error, statistic

PREREQUISITE KNOWLEDGE

You should already know how to:

> calculate the mean, variance and standard deviation from raw or summarised data

> calculate the expectation and variance of a random variable and a linear function of a random variable

> use the normal distribution to calculate probabilities and find values of a random variable.

You should be able to complete the following questions correctly:

1 The ages of people in a cycling club are 17, 25, 21, 32, 40, 28. Calculate the mean and standard deviation of these ages.

2 A set of data is summarised by $n = 50$, $\sum x = 730$, $\sum x^2 = 15126$. Calculate the mean and standard deviation of these data.

3 A discrete random variable has probability distribution given by:

x	1	2	3	4	5	6
$P(X = x)$	0.3	0.15	0.05	0.2	0.25	0.05

Calculate $E(X)$ and $Var(X)$.

4 A random variable X has $E(X) = 11$ and $Var(X) = 4$. Calculate $E(3X + 5)$ and $Var(3X + 5)$.

5 Scores in a test are normally distributed with mean 57 and standard deviation 8. What is the probability of a score being greater than 70?

6 Test scores are normally distributed with mean 65 and standard deviation 10. If no more than 90% of people pass the test, what is the pass mark?

4.1 Populations and samples

A population is a group of items, such as people, animals or objects, that share one or more common features. There are many contexts in which it is necessary to obtain information about the members of a population. A census is an investigation that gathers information about all the members of a population. If the population is large, a census can be very time consuming and costly. In some cases a census is not suitable. In a light-bulb factory, testing all the light bulbs until they failed would mean none were left to be sold. Instead, the information is gathered from a subset of the population, called a sample. The information gathered from the sample is used to make inferences about the population as a whole.

KEY INFORMATION

The population is the entire pool from which a sample is drawn.

An inference is a conclusion reached from looking at evidence.

Example 1

A fundraiser wants to find out if it is a good idea to hold a charity coffee morning in a village hall. She asks 25 people who live in the village if they would attend the coffee morning. Describe:

a the population

b the sample.

Solution

a The population is everyone who lives in the village.

b The sample is the 25 people who were asked.

Example 2

In a factory that manufactures light bulbs, the quality control manager wants to find out how long the light bulbs will last. He selects 100 light bulbs to test. Describe:

a the population

b the sample.

Solution

a The population is all the light bulbs the factory produces.

b The sample is the 100 light bulbs that were selected for testing.

Random sampling

Bias is anything that occurs in the sampling process which means the sample is not representative of the whole population. If you wished to find out whether the population of a city enjoyed watching football, asking people leaving a football match would not give a good representation of the views of the population. One way of reducing bias is to use random sampling.

In simple random sampling, every member of the population has an equal chance of being selected. This means every possible sample of a certain size has an equal chance of being chosen. Random number generators are one way of choosing a simple random sample. Every member of the population is numbered and a random number generator is used to select the sample. Any repeats and numbers that are larger than the population size are ignored.

Random numbers could be generated by pulling numbered balls from a bag, or by using a calculator or a computer. There is some debate about whether random number generators are truly random.

Stop and think Does true randomness really exist?

Example 3

Jamal wishes to find out if the 153 students in his year group get too much homework. He asks the 25 students in his class if they get too much homework.

a Give a reason why this would not be a suitable sample.

b Explain how Jamal could use a random number generator to give a suitable sample of 25 students.

Solution

a The teachers who teach Jamal's class might give more or less homework than those who teach other classes in his year group.

b Jamal could number the students in his year group from 001 to 153. He could use a random number generator to generate 3-digit random numbers. He could ignore any random numbers that are greater than 153 and any repeats, until 25 students had been selected.

> **KEY INFORMATION**
>
> To use a random number generator to select a sample:
>
> - number all members of the population, using numbers with the same amount of digits
> - ignore random numbers that are larger than the population size
> - ignore repeats.

Example 4

A list of people who are registered owners of vehicles is used to obtain a sample of people living in a city. Give two reasons why this would not give a good representation of the population of the city.

Solution

It would not include children.

It would not include adults who do not own a vehicle.

Exercise 4.1A

MM 1 Staff at a doctor's surgery wish to find out how many patients do not keep appointments. They note how many of the first 25 patients on a Monday morning do not turn up.

    a Describe the population.

    b Describe the sample.

    c Give a reason why this is not a suitable sample.

MM 2 A factory manufactures batteries. They want to test how long the batteries will last. Explain why they would need to use a sample.

MM 3 A school wishes to find out how its 874 students travel to school. Describe how they could use a random number generator to give a random sample of 50 students to ask.

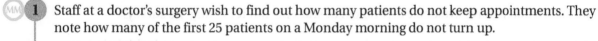

Ⓒ **Communication**   ⓂⓂ **Mathematical modelling**   ⓅⓈ **Problem solving**   73

4 Anya is investigating the cost of train journeys. She decides to ask 20 people who are waiting at a railway station at 7 am how much their tickets cost. Give two reasons why this is not a suitable sample.

5 Give three reasons why using a telephone directory to obtain a sample of people living in a town would not give a good representation of the population.

6 Paul wants to choose one person at random from Kim, Mike and Nora. He throws two fair dice. If the dice both show an even number, he chooses Kim. If the dice both show an odd number, he chooses Mike. If the dice show one even and one odd number, he chooses Nora.

  a Explain why this is not a fair method for choosing a person at random.

  b Describe how Paul could use the two dice to give a fair method for choosing a person at random.

7 Kris wants to select a random sample of 10 students from his year group. There are 65 students in his year group. He uses a random number generator and obtains the following random numbers.

  603121   722828   535813   387575   925561

  Explain how Kris could use these random numbers to select his sample.

8 Read each of these statements, then say whether it is always true, sometimes true, or never true.

  a A sample is a subset of the population.

  b A random sample is generated using random numbers.

  c Using a sample is better than carrying out a census.

  d Using a random sample means there is no bias.

9 Jo wants to find out if the students in her school would be interested in reading a school newspaper. She randomly selects 50 students and sends them a survey to fill in. 30 students return the survey and 22 of these say they would be interested in reading the newspaper. Jo says: "The majority of students are interested in reading a school newspaper."

  a Explain why Jo is wrong to make this conclusion.

  b What should Jo do next to make sure her sample is representative of the whole school?

  c How could Jo have avoided the problem of students not returning the survey?

10 Milun works for a pet food company. He wants to find out how many people think their pet dogs prefer dried food to wet food. Milun includes a survey in randomly selected packets of dried food. He asks owners to fill in the survey and return it to the company.

  a Give two reasons why Milun's method will not give a true random sample.

  b Explain how Milun could improve his method to make the results of the survey more representative.

> **Stop and think**   Investigate Benford's Law of First Digits. Does this support or contradict the idea of using real-life data sets to generate random numbers?

4.2 The mean of sample data

A number that summarises information about a population, such as the mean, is called a population parameter. If such a number is calculated from a sample, it is called a statistic.

The Greek letters μ and σ are used to represent the mean and standard deviation of a population. The sample mean and standard deviation are represented by $\bar{x}$ and s.

If $x_1, x_2, x_3 \ldots$ are the observations from a sample of size n, then the sample mean $\bar{x}$ is given by:

$$\bar{x} = \frac{1}{n} \sum_{i=1}^{n} x_i$$

This can be written more simply as $\bar{x} = \dfrac{\sum x}{n}$.

KEY INFORMATION

$$\bar{x} = \frac{\sum x}{n}$$

Example 5

A random sample of 20 values of a variable X is taken. It is found that $\sum x = 75$. Calculate $\bar{x}$.

Solution

$$\bar{x} = \frac{\sum x}{n}$$

$$\bar{x} = \frac{75}{20}$$

$$= 3.75$$

Example 6

Jan wants to find out how long people spend in her coffee shop. She writes down the time, to the nearest minute, that 10 randomly selected people spend in her shop on a particular day.

Jan records the following results.

31, 20, 52, 46, 28, 39, 37, 43, 17, 48

Calculate the sample mean of Jan's data.

Solution

$$\bar{x} = \frac{\sum x}{n}$$

$$\sum x = 31 + 20 + 52 + 46 + 28 + 39 + 37 + 43 + 17 + 48 = 361$$

$$\bar{x} = \frac{361}{10} = 36.1$$

The sample mean as a random variable

Recall that a random variable is a number obtained as the result of a statistical experiment. As the sample mean is obtained as the result of an experiment (selecting a sample for observations) and may vary between different samples, it is considered to be a random variable.

If X_1, X_2, X_3... are observations from a population with mean μ and variance σ^2, then each observation X_i has expectation μ and variance σ^2.

> Remember that uppercase letters are used to represent random variables, and lowercase letters are used to represent specific values of a random variable.

$$E(\bar{X}) = E\left(\frac{X_1 + X_2 + X_3 + ... + X_n}{n}\right)$$

$$= E\left(\frac{X_1}{n}\right) + E\left(\frac{X_2}{n}\right) + E\left(\frac{X_3}{n}\right) + ... + E\left(\frac{X_n}{n}\right)$$

$$= \frac{1}{n}\left(E(X_1) + E(X_2) + E(X_3) + ... + E(X_n)\right)$$

$$= \frac{1}{n}\left(\mu + \mu + \mu + ... + \mu\right)$$

$$= \frac{1}{n}(n\mu)$$

$$= \mu$$

> These derivations use the properties of linear combinations of random variables.

$$Var(\bar{X}) = Var\left(\frac{X_1 + X_2 + X_3 + ... + X_n}{n}\right)$$

$$= Var\left(\frac{X_1}{n}\right) + Var\left(\frac{X_2}{n}\right) + Var\left(\frac{X_3}{n}\right) + ... + Var\left(\frac{X_n}{n}\right)$$

$$= \frac{1}{n^2}\left(Var(X_1) + Var(X_2) + Var(X_3) + ... + Var(X_n)\right)$$

$$= \frac{1}{n^2}\left(\sigma^2 + \sigma^2 + \sigma^2 + ... + \sigma^2\right)$$

$$= \frac{1}{n^2}(n\sigma^2)$$

$$= \frac{\sigma^2}{n}$$

KEY INFORMATION

If $\bar{X}$ is the sample mean from a sample of size n, taken from a population with mean μ and variance σ^2, then $E(\bar{X}) = \mu$ and

$$Var(\bar{X}) = \frac{\sigma^2}{n}$$

> A larger sample size produces a smaller variance.

Example 7

A random variable X has mean 30.2 and standard deviation 6.1.

a Calculate the expectation and variance of the mean, $\bar{X}$, of a random sample of 20 values of X.

b Calculate the expectation and variance of $5\bar{X} - 2$.

Solution

a $\text{E}(\bar{X}) = 30.2$

$$\text{Var}(\bar{X}) = \frac{6.1^2}{20}$$

$$= 1.86 \text{ (3 s.f.)}$$

b $\text{E}(5\bar{X} - 2) = 5\text{E}(\bar{X}) - 2$

$$= 5 \times 30.2 - 2$$

$$= 149 \text{ (3 s.f.)}$$

$\text{Var}(5\bar{X} - 2) = 5^2\text{Var}(\bar{X})$

$$= 25 \times 1.8605$$

$$= 46.5 \text{ (3 s.f.)}$$

Example 8

The length, in mm, of leaves from an oak tree is known to have mean 86.2 mm and standard deviation 27.6 mm. A sample of 15 leaves is to be taken.

a Calculate the expectation and variance of the sample mean, to 3 s.f.

b Calculate the expectation and variance of the total length of the 15 leaves, to 3 s.f.

Solution

a $\text{E}(\bar{X}) = 86.2$

$$\text{Var}(\bar{X}) = \frac{27.6^2}{15} = 50.8$$

b $\text{E}(15\bar{X}) = 15\text{E}(\bar{X})$

$$= 15 \times 86.2$$

$$= 1290 \text{ (3 s.f.)}$$

$\text{Var}(15\bar{X}) = 15^2\text{Var}(\bar{X})$

$$= 225 \times 50.784$$

$$= 11\,400 \text{ (3 s.f.)}$$

Exercise 4.2A

C 1 A random variable X has mean 6 and standard deviation 2. Calculate the expectation and variance of the mean of a sample of 10 values of X.

C 2 The masses of apples, in grams, are known to have mean 96.3 and standard deviation 5.2. If a sample of 25 apples is to be taken, calculate the expectation and variance of the sample mean.

C 3 At a seaside resort, the number of hours of sunshine per day in July is known to have mean 6.2 and standard deviation 2.5. The hours of sunshine are measured on 10 days in July one year. Calculate the expectation and variance of the sample mean.

C 4 The mean of a random sample of 15 observations of a variable Y is given by $\bar{Y}$. $\bar{Y}$ has expectation 15.2 and variance 9.61. Find the mean and standard deviation of Y.

PS 5 A random variable X has mean 23.2 and standard deviation 4.5. If $\bar{X}$ is the mean of a random sample of 20 observations of X, find the expectation and variance of $3\bar{X} + 2$.

PS 6 The ages, in years, of applicants for jobs at a company have mean 42 and standard deviation 7. If 10 people apply for a particular job, calculate the expectation and variance of the total of their ages.

PS 7 The length of life, in hours, of light bulbs has mean 450 and standard deviation 20. If John buys a pack of 6 bulbs, calculate the expectation and variance of the total lifetime of the bulbs.

PS 8 The cost of producing badges in a particular design comprises a fixed charge of $5 plus $0.20 per badge. The number of badges ordered in each design in a month has mean 613 and standard deviation 51.

    a Find the mean and standard deviation of the monthly cost of producing badges in a particular design.

    b Four designs are chosen at random. Find the expectation and variance of the mean number of badges ordered in each of these designs in a month.

    c Find the expectation and variance of the total monthly cost of producing these four designs.

PS 9 The number of tomatoes produced each day on a farm has mean 430 and standard deviation 15. The profit made by selling each tomato is $0.15.

    a Calculate the mean and standard deviation of the profit made each day on all the tomatoes that are produced.

    b The numbers of tomatoes produced on 10 randomly selected days are recorded. Calculate the expectation and variance of the sample mean.

    c Calculate the expectation and variance of the mean profit made on these 10 days.

PS 10 To hire a car, there is a fixed charge of $75 plus a charge of $50 per day. The number of days for which people hire a car has mean 7.3 and standard deviation 2.1.

    a Calculate the mean and standard deviation of the amount people pay when hiring a car.

    b If 20 people who hire cars are chosen at random, calculate the expectation and variance of the amount each person pays.

PS 11 Employees at a company are paid $60 per day, plus $5 for each complete year of experience they have. The number of years' experience the employees in the company have has mean 6.5 and variance 3.5. Calculate the mean and standard deviation of the total amount paid by the company to five randomly selected employees.

PS 12 The cost of producing advertising posters is $4 per poster, plus a fixed charge of $20. The number of posters in an order has mean 250 and standard deviation 40. Eight orders for posters are chosen at random. Calculate the expectation and variance of the total cost of the 8 orders.

C 13 It is known that for a random sample of 10 values of a variable X, $E(12\bar{X}) = 48.7$ and $\text{Var}(12\bar{X}) = 4.2$. Find μ and σ.

C 14 It is known that for a random sample of 3 values of a variable X, $E(2\bar{X} + 1) = 5.6$ and $\text{Var}(2\bar{X} + 1) = 1.2$. Find μ and σ.

4.3 The probability distribution of the sample mean

As the sample mean is a random variable, it will have a probability distribution. This distribution will give all the values of the sample mean and the probability that each value would occur. The probability distribution of a statistic is known as a sampling distribution.

If a population has a normal distribution, the sample mean will also be normally distributed.

KEY INFORMATION

If $X \sim N(\mu, \sigma^2)$ then

$$\bar{X} \sim N(\mu, \frac{\sigma^2}{n})$$

The proof of this is beyond the mathematics covered in this module.

Example 9

A marine biologist is investigating the masses of grey seals in the Baltic Sea. She knows the mass of female grey seals is normally distributed with mean 145 kg and standard deviation 15.3 kg. She conducts an experiment and captures and weighs 40 adult female seals.

a State the sampling distribution of the mean mass of the 40 seals.

b Calculate the probability that the mean mass of the 40 seals is less than 150 kg.

Solution

a $\bar{X} \sim \text{N}\left(145, \dfrac{15.3^2}{40}\right)$

$\bar{X} \sim \text{N}(145, 5.852)$

b $\text{P}(\bar{X} < 150) = \text{P}(Z < z)$

$z = \dfrac{150 - 145}{\sqrt{5.852}}$

$z = 2.067$

$\text{P}(Z < 2.067) = 0.981$ (3 s.f.) •————

> Use the normal distribution table to find $\text{P}(Z < 2.067)$.

Example 10

A random variable X is normally distributed with mean 40 and standard deviation σ. Random samples of 10 observations of X are taken. It is found that the probability of the mean of these observations being greater than 42 is 0.1.

Calculate σ.

Solution

$\bar{X} \sim \text{N}\left(42, \dfrac{\sigma^2}{10}\right)$

$\text{P}(\bar{X} > 42) = \text{P}(Z > z) = 1 - \text{P}(Z < z) = 0.1$

$\text{P}(Z < z) = 0.9$

$z = 1.282$

$1.282 = \dfrac{42 - 40}{\left(\dfrac{\sigma}{\sqrt{10}}\right)}$

$\sigma = 4.93$ (3 s.f.)

> **KEY INFORMATION**
>
> If X_1, X_2, X_3… are a random sample from any distribution with mean μ and variance σ^2, and the sample is large enough, then the sample mean has approximate distribution
> $$\bar{X} \sim \text{N}\left(\mu, \dfrac{\sigma^2}{n}\right)$$

The Central Limit Theorem

The Central Limit Theorem tells you that the sample mean has an approximate normal distribution, even if the population does not have a normal distribution. The Central Limit Theorem can only be used if the sample is large enough. A sample size of 30 or more is considered large enough. If you know that the population has a normal distribution, you do not need to use the Central Limit Theorem.

> **KEY INFORMATION**
>
> Use the Central Limit Theorem if:
>
> ▸ you do not know the distribution of the population or you know the distribution of the population is not normal
>
> ▸ the sample size is 30 or more.

> **Stop and think** Investigate why a sample size of 30 or more is considered to be a large enough sample to use the Central Limit Theorem.

Example 11

The breaking weight of ropes is known to be normally distributed. Would you need to use the Central Limit Theorem to decide whether the mean breaking weight of a sample of 20 ropes is normally distributed? Give a reason for your answer.

Solution

You would not need to use the Central Limit Theorem. The population of the breaking weight of all the ropes is known to be normally distributed. Therefore the mean of the sample is normally distributed.

Example 12

The value of cars parked in a car park has an unknown distribution. Would you be able to use the Central Limit Theorem to decide whether the mean value of a sample of cars is normally distributed for a sample of:

a 10 cars   **b** 50 cars?

Solution

a No, as the sample size is not large enough. It is smaller than 30.

b Yes, as the sample size is large enough. It is bigger than 30.

Example 13

$X \sim B(12, 0.4)$. A random sample of 50 observations of X is taken.

Find the probability that the mean of these observations is less than 5.

Solution

$E(X) = np$

$\quad = 4.8$

$Var(X) = np(1 - p)$

$\quad = 2.88$

$\bar{X} \sim N\left(4.8, \dfrac{2.88}{50}\right)$ ⟵ Using the Central Limit Theorem.

$z = \dfrac{5 - 4.8}{\left(\sqrt{\dfrac{2.88}{50}}\right)}$

$z = 0.833$

$P(Z < 0.833) = 0.798$ (3 s.f.)

Exercise 4.3A

(PS) 1 The speeds of cars, in $km\,h^{-1}$, passing a police speed trap are normally distributed with mean 35 and standard deviation 5. A random sample of the speeds of 40 cars is taken.

    a State the sampling distribution of the mean speed of the 40 cars.

    b Calculate the probability that the mean speed of the 40 cars is less than $36\,km\,h^{-1}$.

(PS) 2 The heights of 12-year-old boys are normally distributed with mean 151.2 cm and standard deviation 6.3 cm. A random sample of the heights of 100 boys is taken.

    a State the sampling distribution of the mean height of the 100 boys.

    b Calculate the probability that the mean height of the 100 boys is greater than 153 cm.

(C) 3 The diameters of bolts produced by a machine have mean 8.2 mm and standard deviation 0.3 mm. A random sample of 100 bolts is taken.

    a Calculate the expectation and variance of the sample mean.

    b State the sampling distribution of the sample mean.

    c Explain, with a reason, whether you needed to use the Central Limit Theorem in your answer to part **b**.

(PS) 4 A factory produces boxes of matches. The number of matches in a box has mean 52 and standard deviation 4. A random sample of 120 boxes is taken.

    a Calculate the probability that the mean number of matches in the 120 boxes is less than 51.

    b Explain, with a reason, whether you needed to use the Central Limit Theorem in your answer to part **a**.

(PS) 5 The mass of packets of tea are normally distributed with mean 352 g and standard deviation σ. Random samples of 50 packets of tea are taken. It is found that there is a probability of 0.33 that the mean mass of these samples is less than 350 g.

    a Calculate σ.

    b Explain, with a reason, whether you needed to use the Central Limit Theorem in your answer to part **a.**

(PS) 6 The amount of milk, in litres, produced each day by cows on a farm is normally distributed with mean μ and standard deviation 6. Random samples of 20 cows are chosen and the amounts of milk they produce in a day are recorded. There is a probability of 0.25 that the mean amount of milk produced by a sample of 20 cows is greater than 30 litres.

    a Calculate μ.

    b Explain, with a reason, whether you needed to use the Central Limit Theorem in your answer to part **a.**

(PS) 7 In an aptitude test, the mean mark is 74.5 and the standard deviation is 6. A random sample of 100 test marks is taken. Find the probability that the sample mean is within one mark of the population mean.

8 $X \sim \text{Po}(5)$. Random samples of n observations of X are taken. Approximately 1% of the sample means are greater than 6. Estimate n. (You may assume n is large enough for the Central Limit Theorem to apply.)

9 X is a random variable with standard deviation 8.2. Random samples of 100 observations of X are taken. Calculate the probability that the sample mean will differ from the population mean by less than 0.2.

10 Three green counters and two yellow counters are placed in a bag. A counter is drawn at random and not replaced. This is repeated until a yellow counter is drawn. The random variable X is the number of green counters drawn before a yellow counter is drawn.

   a Find the probability distribution of X.

   b Calculate $E(X)$ and $\text{Var}(X)$.

   c A random sample of 50 observations of X is taken. Find the probability that the sample mean is greater than 1.25.

4.4 Unbiased estimates of population mean and variance from sample data

A statistic gives an **unbiased** estimate for a population parameter if the expectation of the statistic is equal to the parameter.

You have already seen that $E(\bar{X}) = \mu$ for a normal distribution or for other distributions if the sample size is large enough. Therefore the sample mean is an unbiased estimate of the population mean.

If $s^2 = \dfrac{1}{n}\displaystyle\sum_{i=1}^{n}(x_i - \bar{x})^2$ is used for the sample variance, $E(s^2) \neq \sigma^2$, and s^2

is therefore not an unbiased estimate of the population variance.

However, if $s^2 = \dfrac{1}{n-1}\displaystyle\sum_{i=1}^{n}(x_i - \bar{x})^2$ is used, then $E(s^2) = \sigma^2$, so s^2 does

give an unbiased estimate of the population variance.

> **Stop and think**
>
> The above result can be proved using algebra, although the proof is rather lengthy. The use of $n - 1$ rather than n is known as Bessel's Correction.
>
> Can you prove that Bessel's Correction gives an unbiased estimate of the population variance?

> **Stop and think**
>
> Can you prove algebraically that the two variance formulae are equivalent?

KEY INFORMATION

Unbiased estimates of the population mean and variance from a sample are given by:

$$\bar{x} = \frac{1}{n}\sum_{i=1}^{n} x_i \text{ often written as}$$

$$\bar{x} = \frac{\sum x}{n}$$

and $s^2 = \dfrac{1}{n-1}\displaystyle\sum_{i=1}^{n}(x_i - \bar{x})^2$

often written as:

$$s^2 = \frac{1}{n-1}\sum(x - \bar{x})^2$$

$$= \frac{1}{n-1}\left(\sum x^2 - \frac{\left(\sum x\right)^2}{n}\right)$$

Example 14

Calculate unbiased estimates of the mean and variance of the population from which the following sample was taken.

6.1, 5.1, 7.2, 8.0, 4.6, 7.2, 6.3, 5.3, 8.1, 4.4

Solution

$$\bar{x} = \frac{\sum x}{n}$$

$$= \frac{62.3}{10}$$

$$= 6.23$$

$$s^2 = \frac{1}{n-1}\left(\sum x^2 - \frac{\left(\sum x\right)^2}{n}\right)$$

$$s^2 = \frac{1}{10-1}\left(404.81 - \frac{62.3^2}{10}\right)$$

$$= 1.85 \text{ (3 s.f.)}$$

Example 15

A forester measures the height, x m, of a sample of 50 trees in a forest. He records the following summary statistics.

$$\sum x = 1762, \sum x^2 = 68\,376$$

Calculate unbiased estimates of the population mean and variance.

Solution

$$\bar{x} = \frac{\sum x}{n} = \frac{1762}{50} = 35.2$$

$$s^2 = \frac{1}{n-1}\left(\sum x^2 - \frac{\left(\sum x\right)^2}{n}\right)$$

$$s^2 = \frac{1}{50-1}\left(68\,376 - \frac{1762^2}{50}\right) = 128 \text{ (3 s.f.)}$$

Example 16

A sample of size 50 is taken from a population and the following summary statistics are obtained.

$$\sum x = 745, \; \sum x^2 = 16\,736$$

a Calculate unbiased estimates of the population mean and variance.

b A sample of size 100 is taken from the same population. Use your answers to part **a** to find the probability that the mean of this sample is greater than 16.1.

Solution

a $\bar{x} = \dfrac{\sum x}{n}$

$= \dfrac{745}{50}$

$= 14.9 \; (3 \text{ s.f.})$

$s^2 = \dfrac{1}{n-1}\left(\sum x^2 - \dfrac{\left(\sum x\right)^2}{n} \right)$

$s^2 = \dfrac{1}{50-1}\left(16\,736 - \dfrac{745^2}{50} \right)$

$= 115$

b $\bar{X} \sim N\left(14.9, \dfrac{115}{100} \right)$

$P\left(\bar{X} > 16.1\right) = P\left(Z > z\right)$

$z = \dfrac{16.1 - 14.9}{\sqrt{\dfrac{115}{100}}} = 1.119$

$P(Z > 1.119) = 1 - 0.8684 = 0.132 \; (3 \text{ s.f.})$

Exercise 4.4A

© **1** Find unbiased estimates of the mean and variance of the population from which the following sample was taken.

35, 37, 42, 41, 34, 45, 39, 37

© **2** Find unbiased estimates of the mean and standard deviation of the population from which the following sample summary statistics were obtained.

$$\sum x = 110, \; \sum x^2 = 900, \; n = 20$$

(C) **3** The number of litres of petrol, x, purchased by 50 customers at a fuel station is recorded. The following summary statistics are noted.

$$\sum x = 1340, \sum x^2 = 36\,296$$

Calculate unbiased estimates of the population mean and variance.

(C) **4** Cartons of milk should each contain 1 litre. A random sample of 10 cartons was taken and the cartons were found to contain the following amounts of milk, in litres:

1.010, 1.005, 1.013, 1.002, 1.009, 1.005, 1.006, 1.011, 1.007, 1.004

Calculate unbiased estimates of the population mean and variance.

(PS) **5** A random sample of 100 observations from a normally distributed population gave the following results.

$$\sum x = 2524, \sum x^2 = 65\,513$$

a Find unbiased estimates of the population mean and standard deviation.

b Use your answers to part **a** to calculate the probability that the mean of a random sample of 40 observations of the population is less than 27.

(C) **6** The number of minutes, x, that a bus is late is recorded for a random sample of 150 buses. The following summary statistics are obtained.

$$\sum x = 765, \sum x^2 = 4712$$

a Calculate unbiased estimates of the population mean and variance.

b Use your answers to part **a** to calculate the probability that the mean number of minutes late of a random sample of 40 buses is less than 6 minutes.

c Explain, with a reason, whether it was necessary to use the Central Limit Theorem to work out your answer to part **b**.

(C) **7** The ages of hospital patients are known to be normally distributed. The ages, x years, of a random sample of 300 hospital patients are recorded, giving the following summary statistics.

$$\sum x = 14\,307, \sum x^2 = 713\,578$$

a Calculate unbiased estimates of the population mean and variance.

b Use the values obtained in part **a** to calculate the probability that the mean age of a random sample of 25 patients is more than 48 years.

c Explain why you did not need to use the Central Limit Theorem in part **b**.

(PS) **8** The scores of 100 students in an exam were recorded and the following summary statistics were obtained:

$$\sum x = 3020, \sum x^2 = 106724$$

a Calculate unbiased estimates of the population mean and variance.

b One of the questions in the exam was found to be impossible to answer and 5 marks were added to each student's score. Estimate the population mean and variance of the new scores.

c A teacher wants to compare the students' marks with those in a previous exam. After the extra 5 marks are added, she doubles each score. Estimate the population mean and variance of the new scores.

PS 9 A large company wants to analyse how many working days employees take off for illness. They take a random sample of 40 employees and find that the number of days off in a year can be summarised by the following summary statistics.

$$\sum x = 136, \ \sum x^2 = 503$$

a Calculate unbiased estimates of the population mean and variance.

b Use your answers to part **a** to find the smallest possible sample size where there is a probability of 0.01 that the sample mean is greater than 3.7. (You may assume the sample size is large enough for the Central Limit Theorem to apply.)

PS 10 a Find unbiased estimates of the mean and standard deviation of the population from which the following summary statistics were taken.

$$\sum x = 2102, \ \sum x^2 = 56\,314, \ n = 80$$

b For a sample size of 200, there is a probability of 0.05 that the sample mean exceeds a particular value. Use your answers to part **a** to find this value.

Mathematics in life and work: Group discussion

You are an account manager at an advertising company. One of your clients is a company that produces light bulbs. You want to base your new campaign on your client's light bulbs lasting longer than those of their main competitor.

1 Why might it be important to be able to back up your campaign with statistics?

2 How could you go about investigating this? What problems might you face?

3 Which statistics would be useful in this investigation? Could you use these to draw inferences about all the light bulbs produced?

4 What values of these statistics might convince you that your campaign is based in fact?

4.5 Confidence interval for a population mean

A confidence interval is a range of values that is likely to contain a population parameter. It is calculated from sample data. The confidence level of an interval is the probability that the population parameter falls within that interval. It is given as a percentage.

A confidence interval is calculated from the point estimate (the statistic obtained from the sample) plus or minus a margin of error. This margin of error is based on the standard deviation (in this case known as the standard error) of the statistic.

If the sample mean is normally distributed, you can use the normal distribution to calculate the confidence interval for a specified confidence level.

For a population with mean μ and variance σ^2, a confidence interval for μ is given by:

$$\bar{x} - \frac{\sigma z}{\sqrt{n}} \leqslant \mu \leqslant \bar{x} + \frac{\sigma z}{\sqrt{n}}$$

where $\bar{x}$ is the sample mean, n is the sample size and z is a value of the random variable $Z \sim N(0, 1^2)$.

The value of z is calculated from the required confidence level. For example, if the required confidence level is 98%, you would use the normal distribution as follows:

You require the shaded area to be 0.98. By the symmetry of the normal distribution, the unshaded areas are both 0.01. Therefore, you require z such that $P(Z < z) = 0.99$. Using the table of critical values of the normal distribution gives $z = 2.326$.

If the population variance is unknown and the sample size is 30 or more, you can use the unbiased estimate of population variance s^2 to calculate the confidence interval.

Remember that if the population is normally distributed, or the Central Limit Theorem can be applied, the sample mean is normally distributed.

Note how this relates to the formula to standardise a normal distribution, $z = \frac{x - \mu}{\sigma}$.

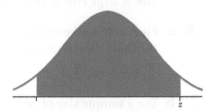

The case where the population variance is unknown and the sample size is small will not be dealt with in this module.

Example 17

The mass, in g, of packets of flour are distributed with mean μ and standard deviation 17. A random sample of 200 packets is taken and found to have mean mass 248 g. Calculate a 99% confidence interval for μ.

Solution

The area of the shaded region must be 0.99. Therefore the unshaded areas are both 0.005 and you require $P(Z < z) = 0.995$.

From the table of critical values of the normal distribution, $z = 2.576$.

$$\bar{x} - \frac{\sigma z}{\sqrt{n}} \leqslant \mu \leqslant \bar{x} + \frac{\sigma z}{\sqrt{n}}$$

$$248 - \frac{17 \times 2.576}{\sqrt{200}} \leqslant \mu \leqslant 248 + \frac{17 \times 2.576}{\sqrt{200}}$$

$245 \leqslant \mu \leqslant 251$ (3 s.f.)

This could also be written as [245, 251].

KEY INFORMATION

$$\bar{x} - \frac{\sigma z}{\sqrt{n}} \leqslant \mu \leqslant \bar{x} + \frac{\sigma z}{\sqrt{n}}$$

Example 18

A random sample of 50 bags of sugar is taken from a production line and found to have a mean mass of 502 g and standard deviation 3.2 g.

a Calculate an unbiased estimate of the standard deviation of all bags of sugar that are produced.

b Calculate a 99% confidence interval for the mean of all bags of sugar that are produced. Explain whether you needed to use the Central Limit Theorem in this calculation.

c Three different random samples of 50 bags of sugar are taken and a 99% confidence interval is calculated for each. Find the probability that none of these intervals contains the population mean.

Solution

a Bessel's Correction needs to be applied to the sample standard deviation to give an unbiased estimate for the population standard deviation.

$$s = \sqrt{\frac{50}{49} \times 3.2^2} = 3.23$$

b $\mathrm{P}(Z < z) = 0.995$

$$z = 2.576$$

$$502 - \frac{3.23 \times 2.576}{\sqrt{50}} \leqslant \mu \leqslant 502 + \frac{3.23 \times 2.576}{\sqrt{50}}$$

$500.8 \leqslant \mu \leqslant 503.2$ (to 1 d.p.)

The Central Limit Theorem was needed in this calculation as the distribution of the population was unknown.

c The probability that a confidence interval does not contain the mean is $1 - 0.99 = 0.01$. The events of two or more confidence intervals containing the mean are independent.

$$0.01^3 = 0.000001$$

Exercise 4.5A

C 1 The masses of a species of hamsters are known to be normally distributed with standard deviation 3.2 g. A random sample of 20 hamsters are weighed and found to have mean mass 24.1 g. Calculate a 95% confidence interval for the population mean.

C 2 The lengths of climbing ropes produced in a factory are normally distributed with standard deviation 0.3 cm. A random sample of 25 ropes is measured and found to have mean length 501 cm. Calculate a 98% confidence interval for the population mean.

(C) 3 The masses, in grams, of eggs produced by hens at a free range farm are normally distributed with mean μ and variance σ^2. A random sample of 125 eggs are weighed and found to have mean mass 58 g and standard deviation 5 g. Calculate a 98% confidence interval for μ.

(PS) 4 A teacher wishes to investigate how long her students spend on their maths homework. She asks a random sample of 50 students to record the number of minutes, x, that they spend on their maths homework in a week. The following statistics are obtained.

$$\sum x = 3012, \sum x^2 = 189\,354$$

a Calculate a 99% confidence interval for the mean number of minutes that all her students spend on their homework.

b Explain whether you needed to use the Central Limit Theorem in this calculation.

(PS) 5 The masses of raspberries, in grams, grown on a fruit farm are normally distributed with standard deviation 0.5 g. A random sample of 100 raspberries is found to have mean mass 4.2 g.

a Calculate a 97% confidence interval for the population mean.

b Three different random samples of 100 raspberries are taken and a 97% confidence interval calculated for each. Find the probability that none of these intervals contains the true value of the population mean.

(PS) 6 Blood cholesterol levels, in mmol/litre, of patients at a doctor's surgery are known to be normally distributed with standard deviation 0.6. The doctor wishes to calculate a 95% confidence interval for the population mean, with total width less than 0.8. Calculate the smallest sample size that he requires.

(PS) 7 A random sample of 50 packets of biscuits produced in a factory are weighed and their masses, m g, are recorded. These are summarised as follows.

$$\sum m = 15\,924, \sum m^2 = 5\,085\,213$$

a Calculate unbiased estimates of the population mean and variance.

b Calculate a 92% confidence interval for the population mean.

c An x% confidence interval for the population mean has total width 10. Find x.

d Four random samples of 10 packets of biscuits are taken and a 90% confidence interval for the population mean is calculated from each sample. Find the probability that at least one of these intervals contains the true value of the population mean.

(PS) 8 The heights of men, in metres, are known to have a normal distribution. A random sample of the heights of 200 men was taken and a 98% confidence interval was calculated to be $179.2 \leqslant \mu \leqslant 182.6$.

a Calculate the value of the sample mean.

b Calculate the value of the population standard deviation.

c Find a 99% confidence interval for the population mean.

d 50 random samples of 200 men are taken and a 98% confidence interval is calculated for each. How many of these samples would you expect to contain the population mean?

(PS) **9** A random variable X is known to have a normal distribution with mean μ and standard deviation 0.5. A random sample of 10 values of X is taken with the following results.

13.2, 11.3, 13.6, 10.3, 12.3, 12.4, 11.2, 10.7, 12.6, 9.6

a Find a 95% confidence interval for μ.

b A random sample of 20 values of X has the same mean. A confidence interval calculated from this sample has the same width as before. Find the confidence level of this interval.

(PS) **10** A random variable X has a standard deviation of 35. It is required to have a confidence level of 95% that the sample mean differs from the true mean by less than 10. Determine the minimum sample size needed.

4.6 Confidence interval for a population proportion

A population proportion is the fraction of a population that have a particular feature, for example, those who own a car. You would describe having this feature as a success.

Consider a random sample of size n taken from a population in which the proportion of successes is p and the proportion of failures is $1 - p$. The number of successes, X, in the sample, has a binomial distribution. The proportion of successes in the sample is given by the random variable $P_s = \dfrac{X}{n}$. As X has $\mathrm{E}(X) = np$ and $\mathrm{Var}(X) = np(1 - p)$, then you can calculate the distribution of P_s using linear combinations of random variables.

> Refer to **Probability & Statistics 1, Chapter 3, Discrete random variables**.

$\mathrm{E}(P_s) = \mathrm{E}\left(\dfrac{X}{n}\right)$

$\quad = \dfrac{1}{n}\mathrm{E}(X)$

$\quad = \dfrac{1}{n}np$

$\quad = p$

$\mathrm{Var}(P_s) = \mathrm{Var}\left(\dfrac{X}{n}\right)$

$\quad = \dfrac{1}{n^2}\mathrm{Var}(X)$

$\quad = \dfrac{1}{n^2}np(1 - p)$

$\quad = \dfrac{p(1 - p)}{n}$

KEY INFORMATION

An approximate confidence interval for a population proportion p is given by:

$$p_s - z\sqrt{\dfrac{p_s(1 - p_s)}{n}} \leqslant p$$

$$\leqslant p_s + z\sqrt{\dfrac{p_s(1 - p_s)}{n}}$$

where p_s is the sample proportion, n is the sample size and z is a value of the random variable $Z \sim \mathrm{N}(0, 1^2)$ that is calculated from the required confidence level.

As $\mathrm{E}(P_s) = p$, then p_s is an unbiased estimator for p, and $\sqrt{\dfrac{p_s(1 - p_s)}{n}}$ gives the standard error. If n is large, P_s is approximately normally distributed. Therefore you can use the standard normal distribution to calculate a confidence interval for p.

You would describe this as an approximate interval as the normal approximation to the binomial distribution is used in its calculation. This means it is only suitable when the sample size is large. Note that for a confidence interval, a continuity correction is not applied.

Example 19

An experiment is performed 100 times. The number of successful outcomes is 46. Find an approximate 99% confidence interval for p, the probability of a successful outcome.

Solution

$p_s = \dfrac{46}{100}$

$\quad = 0.46$

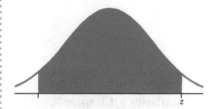

You require the shaded area to be 0.99. Therefore the two unshaded areas are both 0.005 and you require z such that $P(Z < z) = 0.995$. From the critical values of the normal distribution table, $z = 2.576$.

$$p_s - z\sqrt{\dfrac{p_s(1 - p_s)}{n}} \leqslant p \leqslant p_s + z\sqrt{\dfrac{p_s(1 - p_s)}{n}}$$

$$0.46 - 2.576\sqrt{\dfrac{0.46 \times 0.54}{100}} \leqslant p \leqslant 0.46 + 2.576\sqrt{\dfrac{0.46 \times 0.54}{100}}$$

$0.332 \leqslant p \leqslant 0.588$

This interval could also be written as [0.332, 0.588].

Example 20

A random sample of 200 people living in a town is taken. It is found that 86 of them own cars. Calculate an approximate 95% confidence interval for the proportion of people living in the town who own cars.

Solution

$p_s = \dfrac{86}{200}$

$\quad = 0.43$

You require the area of the shaded region to be 0.95. Therefore the two unshaded areas are both 0.025 and you require z such that $P(Z < z) = 0.975$. From the critical values of the normal distribution table, $z = 1.960$.

$$p_s - z\sqrt{\frac{p_s(1 - p_s)}{n}} \leqslant p \leqslant p_s + z\sqrt{\frac{p_s(1 - p_s)}{n}}$$

$$0.43 - 1.960\sqrt{\frac{0.43 \times 0.57}{200}} \leqslant p \leqslant 0.43 + 1.960\sqrt{\frac{0.43 \times 0.57}{200}}$$

$$0.361 \leqslant p \leqslant 0.499$$

This could also be written as [0.361, 0.499].

Exercise 4.6A

(C) 1 An experiment was performed 500 times. The number of successful outcomes was 365. Find an approximate 90% confidence interval for the probability of a successful outcome.

(C) 2 In a random sample of 150 households in a town, 120 households owned at least one laptop computer. Find an approximate 95% confidence interval for the proportion of families in the town who own at least one laptop computer.

(C) 3 In a market research survey, 35 people out of a random sample of 200 in a large city said they regularly used a certain brand of washing powder. Calculate an approximate 97% confidence interval for the proportion of families in the city who regularly use that brand of washing powder.

(C) 4 A die is thrown 200 times and lands on an even number on 58 of these throws. Calculate an approximate 99% confidence interval for the probability that the die lands on an even number on one throw.

(PS) 5 A biased coin is thrown 100 times and 24 throws result in the coin showing tails. A confidence level of $x\%$ for the probability of the coin showing tails gives an approximate confidence interval of width 0.15. Calculate x.

(PS) 6 In a survey, 1000 people were asked and 739 said they preferred salted popcorn to sweet popcorn.

a Find an approximate 98% confidence interval for the proportion of the population who prefer salted popcorn.

b For the same sample, an approximate confidence interval of total width 0.05 has a confidence level of $x\%$. Find x.

c For a different sample of 200 people, the approximate confidence interval is given by $0.388 \leqslant p \leqslant 0.422$. Find the number of people in this sample who preferred salted popcorn.

7 An experiment was carried out 250 times. The number of successful outcomes was 120.

a Calculate an approximate 98% confidence interval for p, the probability of a successful outcome.

b Explain why the confidence interval is described as approximate.

c Describe the effect on the width of the confidence interval if the confidence level is increased.

d Describe the effect on the width of the confidence interval if the number of times the experiment is carried out is increased.

8 In a survey, a random sample of 500 people were asked if they added sugar to their coffee. From the survey, an approximate confidence interval for the proportion of the population who add sugar to their coffee was calculated to be [0.288, 0.372].

a How many people in the sample said they added sugar to their coffee?

b Calculate the confidence level that was used.

9 An experiment was carried out 50 times. The number of successful outcomes was 35. Find the number of times the experiment would have to be carried out so that an approximate 90% confidence interval for p, the probability of success, has width less than 0.1.

10 In a random sample of 50 electrical shops, it was found that 10 of them sold a particular kettle at a price lower than that recommended by the manufacturer. For the proportion of all electrical shops selling the kettle at the lower price, an approximate confidence interval of width 0.2 was calculated.

a Calculate the confidence level that was used.

b Interpret your answer to part **a**.

Mathematics in life and work: Group discussion

You have just been appointed as the new production manager in a light-bulb factory. You decide to investigate how many light bulbs get broken during the packing process.

1 How would you gather data to investigate this?

2 Which statistic would be most useful?

3 Would you be able to use this statistic to make inferences about all the light bulbs that are produced? If so, how confident would you be that you were correct?

4 What value of this statistic would convince you that the number of light bulbs getting broken during the packing process was acceptable?

SUMMARY OF KEY POINTS

› A population is a group of items, such as people, animals or objects, that share one or more common features.

› A sample is a subset of a population from which information is gathered and used to make inferences about the whole population.

› A sampling method is unsuitable if it introduces bias. You should be able to comment on why a particular sampling method is unsuitable.

› The sample mean is given by $\bar{x} = \dfrac{\sum x}{n}$.

› The sample mean is a random variable, $\bar{X}$, with $E(\bar{X}) = \mu$ and $Var(\bar{X}) = \dfrac{\sigma^2}{n}$.

› If the population is normally distributed, $\bar{X} \sim N\left(\mu, \dfrac{\sigma^2}{n}\right)$.

› If the population is not normally distributed, but the sample size is greater than 30, the Central Limit Theorem tells you that $\bar{X} \sim N\left(\mu, \dfrac{\sigma^2}{n}\right)$.

› An unbiased estimate of the population mean is given by $\bar{x} = \dfrac{\sum x}{n}$.

› An unbiased estimate of the population variance is given by

$$s^2 = \frac{1}{n-1}\sum(x - \bar{x})^2 = \frac{1}{n-1}\left(\sum x^2 - \frac{\left(\sum x\right)^2}{n}\right).$$

› A confidence interval for the population mean is given by $\bar{x} - \dfrac{\sigma z}{\sqrt{n}} \leqslant \mu \leqslant \bar{x} + \dfrac{\sigma z}{\sqrt{n}}$. If the sample size is greater than 30 and the population variance is unknown, the unbiased estimate of population variance s^2 may be used to calculate the confidence interval.

› An approximate confidence interval for a population proportion is given by

$$p_s - z\sqrt{\frac{p_s(1 - p_s)}{n}} \leqslant p \leqslant p_s + z\sqrt{\frac{p_s(1 - p_s)}{n}}.$$

EXAM-STYLE QUESTIONS

1 Shreena wishes to find out if it is a good idea to start a hockey club at school. She decides to ask the first 10 people she sees one day if they would like to join a hockey club.

    a Give two reasons why this would not give a representative sample.

    b Explain how Shreena could obtain a random sample of 10 students in her school to ask.

2 The masses, in kg, of babies born in a hospital are distributed with mean μ and standard deviation 0.235. A random sample of 200 babies is found to have mean mass 3.425 kg. Find a 99% confidence interval for μ.

3 The weight loss, in kilograms, of participants in a medical trial during one month is normally distributed with mean 3.5 and standard deviation 0.8. State the distribution of the total weight loss of 5 randomly selected participants during this month.

C 4 **a** A random variable Y has mean μ and standard deviation σ. A random sample of n values of Y has mean $\bar{Y}$. State $E(\bar{Y})$ and $Var(\bar{Y})$ in terms of μ, σ and n.

b Given that Y is normally distributed, state the distribution of $\bar{Y}$.

c Given that Y has mean 35 and standard deviation 4, state the distribution of the mean of a random sample of 10 values of Y.

C 5 The amounts of orange juice, in ml, in a random selection of 5 cartons are given below.

253.2, 255.1, 259.3, 253.4, 254.7

Calculate unbiased estimates of the population mean and standard deviation.

C 6 In a survey of 200 random people in a city, 160 are pet owners. Find an approximate 95% confidence interval for the proportion of people in the city who own a pet.

C 7 A random sample of 50 observations of a random variable X is taken. The results are shown below.

$$\sum x = 113.4, \sum x^2 = 305.4$$

a Calculate unbiased estimates of the population mean and variance of X.

b Hence calculate a 98% confidence level for the population mean.

c State, with a reason, whether you needed to use the Central Limit Theorem in part **b**.

C 8 The scores in 50 randomly selected rounds of golf played on a course are given below.

Score (x)	69	70	71	72	73	74
Frequency (f)	3	8	12	18	7	2

a Calculate unbiased estimates of the population mean and variance for all rounds of golf played at the course.

b Use your answers to part **a** to find a 96% confidence interval for the population mean.

c Explain what is meant by a 96% confidence interval for the population mean.

d Three different random samples of 50 rounds of golf are taken and a 96% confidence interval is calculated for each. Find the probability that all three intervals will contain the population mean.

PS 9 The length of time, in minutes, that cars spend in a car park is normally distributed with mean 125 and standard deviation 20.

a A random sample of 50 cars is taken. Find the probability that the sample mean is greater than 131 minutes.

b A random sample of 100 cars is taken. Find the probability that the sample mean is between 120 and 130 minutes.

c Explain whether you would still have been able to calculate these probabilities if the population was not normally distributed.

C 10 A normally distributed random variable X has mean μ and standard deviation σ. The mean of 20 randomly selected observations of X is given by $\bar{X}$.

   a Given that $E(\bar{X}) = 12$, find μ.

   b Given that $Var(\bar{X}) = 3.5$, find σ.

   c 100 such samples of 20 observations are taken. In how many of these would you expect the sample mean to be greater than 12.5?

   d Explain whether the Central Limit Theorem was needed when calculating your answer to part **c**.

PS 11 In a survey, 500 people were asked and 123 said they preferred decaffeinated coffee to normal coffee.

   a Find an approximate 95% confidence interval for p, the proportion of the population who prefer decaffeinated coffee.

   b For the same sample, an approximate confidence interval of total width 0.1 has a confidence level of x%. Find x.

   c For a different sample of 100 people, an approximate confidence interval is given by $0.22 \leqslant p \leqslant 0.28$. Find the number of people in this sample who preferred decaffeinated coffee.

PS 12 The dividends paid to investors in a company in one year are normally distributed with mean $520 and standard deviation σ. A random sample of 25 investors is chosen. It is found that there is a probability of 0.2 that the mean amount of dividends paid to these investors is more than $550. Calculate σ.

PS 13 The heights of trees, in metres, in a forest are normally distributed with mean μ and standard deviation σ. A sample of 50 trees is taken and a 98% confidence interval for μ is calculated and found to be [35.2, 39.6].

   a Calculate the value of the sample mean.

   b Find the value of σ.

   c Find a 99% confidence interval for μ.

   d Find the smallest sample size required to give a confidence interval with width less than 3 and a confidence level of 90%.

MM 14 A political campaigner wishes to find out what proportion of the voting population in the electoral district will vote for her candidate.

   a Give a reason why asking 200 people leaving a campaign meeting would not give a representative sample.

   b Explain how she could generate a random sample of 200 voters to ask.

   c She finds that 150 of the 200 voters she asks plan to vote for her candidate. Calculate an approximate 95% confidence interval for the proportion of the population that plan to vote for her candidate.

d She asks three separate random samples of 200 voters if they plan to vote for her candidate. An approximate 95% confidence interval is calculated for each sample. Find the probability that at least one of the confidence intervals contains the true population proportion.

e She wishes to achieve a confidence level of 99% with an approximate confidence interval of width less than 0.1. She uses the proportion of voters who would support her candidate in the original sample of 200 as an unbiased estimator for the population proportion. What is the smallest sample size she would need to use to achieve the required confidence level and interval?

C **15** The random variable X is the number of tails obtained when 3 fair coins are thrown.

a Find the probability distribution of X.

b Calculate $E(X)$ and $Var(X)$.

c A random sample of 40 observations of X is taken. Find the probability that the sample mean is less than 1.9.

d Justify your use of the normal distribution in part **c**.

C **16** $X \sim B(10, 0.35)$. A random sample of 100 observations of X is taken.

a Find the probability that the mean of these observations is greater than 3.8.

b Explain whether you needed to use the Central Limit Theorem in your answer to part **a**.

MM **17** **a** Give two reasons why a sample might be preferred to a census and explain what is meant by saying a sample is random.

b Explain how the following random numbers could be used to choose a random sample of six observations from a population with 50 members.

508 833   332 695   128 517   644 017

c A random sample of six observations from a normally distributed population with variance 1.05 is taken, giving the following results.

47   50   49   46   52   51

Use this sample to find a 98% confidence interval for the population mean.

MM **18** In a survey, 150 out of 400 families said they own a dishwasher.

a Find the number of families who would need to be asked so that an approximate 99% confidence interval for the proportion of all families who own a dishwasher has width less than 0.05.

b Explain what is meant by a 99% confidence interval in this context.

PS **19** The random variable X has distribution Po(2). Random samples of 200 observations of X are taken.

a Describe fully the distribution of the sample mean.

b Find the probability that the mean of a random sample of 120 observations of X is greater than 2.2.

c Justify your use of the normal distribution in part **b**.

Mathematics in life and work

You work in a factory which is producing a new type of long-life light bulb. You wish to promote the new bulbs by stating the mean lifetime of the bulbs on the packaging.

1   You take a random sample of 200 light bulbs and calculate the following summary statistics for the lifetime in hours, x.

$$\sum x = 225\,134, \sum x^2 = 255\,432\,150$$

Calculate unbiased estimates for the mean and standard deviation of the lifetimes of all the bulbs of this type.

2   Use your answers to part **a** to find a 98% confidence interval for the mean lifetime of all the bulbs.

3   Give one way in which you could decrease the width of this confidence interval without reducing the confidence level.

4   Using your answer to part **b**, what value of the expected lifetime of a bulb would you put on the packaging? Give a reason for your answer.

5 HYPOTHESIS TESTS

Mathematics in life and work

Hypothesis testing is used in many careers to test whether a supposition is true, for example:

» If you were a driving instructor, you might want to test whether a new teaching method has improved the pass rate of your pupils.

» If you were a road safety officer, you might want to test whether building a bypass around a town has reduced the number of road accidents that occur in the town.

» If you managed a medical clinic, you might want to test whether an automated check-in system has reduced the amount of time patients wait for appointments.

LEARNING OBJECTIVES

You will learn how to:

» understand the nature of a hypothesis test, the difference between one-tailed and two-tailed tests, and the terms null hypothesis, alternative hypothesis, significance level, rejection region (or critical region), acceptance region and test statistic

» formulate hypotheses and carry out a hypothesis test in the context of a single observation from a population which has a binomial or Poisson distribution, using:
  » direct evaluation of probabilities
  » a normal approximation to the binomial or the Poisson distribution, where appropriate

» formulate hypotheses and carry out a hypothesis test concerning the population mean in cases where the population is normally distributed with known variance or where a large sample is used

» understand the terms Type I error and Type II error in relation to hypothesis tests

» calculate the probabilities of making Type I and Type II errors in specific situations involving tests based on a normal distribution or direct evaluation of binomial or Poisson probabilities.

LANGUAGE OF MATHEMATICS

Key words and phrases you will meet in this chapter:

» acceptance region, alternative hypothesis, critical value, null hypothesis, one-tailed test, rejection region, significance level, test statistic, two-tailed test, Type I error, Type II error.

PREREQUISITE KNOWLEDGE

You should already know how to:

- identify, from a context, whether a distribution is binomial or Poisson
- calculate probabilities directly for a binomial or Poisson distribution
- use tables to find probabilities for a normal distribution
- approximate the binomial or Poisson distribution with the normal distribution and know the conditions under which this can be done
- calculate unbiased estimates of the population mean and variance from a sample
- use the fact that the sample mean is normally distributed, with mean μ and variance $\frac{\sigma^2}{n}$, when the population is normally distributed or the sample size is large.

You should already be able to answer these questions:

1   $X \sim B(10, 0.4)$
Find: **a** $P(X \leq 2)$ **b** $P(X \geq 9)$.

2   $X \sim Po(5)$
Find: **a** $P(X \leq 1)$ **b** $P(X \geq 3)$.

3   $X \sim B(50, 0.45)$. Use a suitable approximation to find $P(X \geq 30)$.

4   $X \sim Po(24)$. Use a suitable approximation to find $P(X \leq 17)$.

5   A sample is defined by the following summary statistics.

$$\sum x = 675, \sum x^2 = 20\,132, n = 30$$

Calculate unbiased estimates of the population mean and variance.

6   Identify whether each of these contexts would be modelled with the binomial or Poisson distribution.

   a   A fair die is rolled 50 times.

   b   A call centre receives calls at the rate of 15 per minute.

7   State the conditions under which the Central Limit Theorem is needed and can be used.

5.1 Hypothesis testing

A statistical hypothesis is a statement about the value of a population parameter, which may or may not be true. Hypothesis testing involves using data from a sample to test a hypothesis. The statistic that is used for the hypothesis test is called the test statistic.

Null and alternative hypotheses

A hypothesis test starts with defining null and alternative hypotheses. The null hypothesis states the expected or theoretical value of a population parameter. The alternative hypothesis states that the population parameter has changed or is significantly different from the expected value.

The notation H_0 is used for the null hypothesis and H_1 for the alternative hypothesis.

You could carry out a hypothesis test to find out if a die was equally likely to land on 3 as on any other number. The theoretical probability of a fair die landing on any side is $\frac{1}{6}$. Therefore if the die is fair, the probability of it landing on 3 is $\frac{1}{6}$.

The null hypothesis for this test would be written as H_0: $p = \frac{1}{6}$.

If you thought the die was biased, your alternative hypothesis would state that the probability of it landing on a 3 was something other than $\frac{1}{6}$.

There are three different possibilities for the alternative hypothesis.

If you thought the die was biased towards 3, your alternative hypothesis would be that the die lands on 3 more often than you would expect. This would be written as H_1: $p > \frac{1}{6}$.

If you thought the die was biased against 3, your alternative hypothesis would be that the die lands on 3 less often than you would expect. This would be written as H_1: $p < \frac{1}{6}$.

If you thought the die was biased but were not sure how, your alternative hypothesis would be that the probability of landing on 3 was not $\frac{1}{6}$. This would be written as H_1: $p \neq \frac{1}{6}$.

One-tailed and two-tailed tests

The first two versions of the alternative hypothesis give a one-tailed test. In a one-tailed test the alternative hypothesis states that the parameter is either greater than or less than the value given in the null hypothesis.

If you suspect that the value of the parameter is greater than in H_0, use a one-tailed test and look at the upper tail of the distribution.

If you suspect that the value of the parameter is less than in H_0, use a one-tailed test and look at the lower tail of the distribution.

The third version of the alternative hypothesis gives a two-tailed test. In a two-tailed test, the alternative hypothesis only states that the parameter is not equal to the value given in the null hypothesis.

If you suspect that the value of the parameter has changed but you are not sure how, use a two-tailed test. You can look at either tail in your test. The tail you choose will depend on the value of your test statistic.

KEY INFORMATION

The null hypothesis H_0 states the expected or theoretical value of a population parameter. The alternative hypothesis H_1 states that the population parameter has changed or is different from the expected value.

The one-tailed test gets its name from testing the region under one of the tails (sides) of the distribution.

KEY INFORMATION

In a one-tailed test the alternative hypothesis states that the parameter is either greater than or less than the value given in the null hypothesis. In a two-tailed test, the alternative hypothesis only states that the parameter is not equal to the value given in the null hypothesis.

The two-tailed test gets its name from testing the regions under both of the tails (sides) of the distribution.

Example 1

Sunjit rolls a tetrahedral die, which he thinks is biased towards 1. He wants to do a hypothesis test to test his theory. He rolls the die 40 times and counts the number of times it lands on 1.

a Describe the test statistic, X.

b Write down a suitable null hypothesis.

c Write down a suitable alternative hypothesis.

d Is this a one-tailed or two-tailed test?

Solution

a The test statistic, X, is the number of times the die shows 1 in 40 rolls.

b The null hypothesis gives the theoretical probability of rolling a 1 if the die is unbiased. A tetrahedral die has four sides so the theoretical probability of rolling a 1 is 0.25.

H_0: $p = 0.25$

c The alternative hypothesis is what Sunjit believes is going to happen. Sunjit thinks his die is biased towards 1, so there would be a greater probability of it landing on a 1.

H_1: $p > 0.25$

d The alternative hypothesis specifies that p is greater than 0.25, so the test is one-tailed.

Example 2

Ana wants to find out if a coin is biased. She decides to carry out a hypothesis test. She throws the coin 10 times and records how many times it lands on heads.

a Describe the test statistic, X, and state its distribution.

b Write down suitable null and alternative hypotheses.

c Write down the distribution of the test statistic if your null hypothesis is true.

d Is this a one-tailed or two-tailed test?

Solution

a The test statistic, X, is the number of times the coin lands on heads in 10 throws. There are a fixed number of trials, each of which ends in success (heads) or failure (tails). The probability of success is the same for each trial. This means X has a binomial distribution, $X \sim B(10, p)$, where p is the probability of the coin landing on heads when it is thrown.

b If the coin is not biased, the probability it lands on heads is 0.5. If the coin is biased, the probability it lands on heads is not 0.5.

The null hypothesis would be $H_0: p = 0.5$.

The alternative hypothesis would be $H_1: p \neq 0.5$.

c If the null hypothesis is true, then $X \sim B(10, 0.5)$

d The alternative hypothesis just states that $p \neq 0.5$, so this is a two-tailed test.

Exercise 5.1A

 1 A child can catch a ball with a probability of success of 0.55. His parents practise with him for 10 minutes every day to help him improve. The parents want to see if he can now catch more successfully. They throw the ball to him 20 times and record how often he catches it.

   a Describe the test statistic, X.

   b Write down a suitable null hypothesis.

   c Write down a suitable alternative hypothesis.

   d Is this a one-tailed or two-tailed test?

 2 Hwang runs a cheese stall at a Christmas market. He knows the probability of selling the cheese to a passing customer is 0.68. Hwang decides to change the labels and packing to see if this makes a difference to the number of customers who buy cheese. He records whether the next 50 passing customers buy cheese.

   a Describe the test statistic, X.

   b Write down a suitable null hypothesis.

   c Write down a suitable alternative hypothesis.

   d Is this a one-tailed or two-tailed test?

 3 The probability of Nico winning a computer game is 0.65. Nico practises and believes he is now much better and wants to test his theory. He plays the computer game 10 times and records how many times he wins.

   a Describe the test statistic, X.

   b Write down a suitable null hypothesis.

   c Write down a suitable alternative hypothesis.

   d Is this a one-tailed or two-tailed test?

 4 Monisha is a road safety officer. Over a long period of time, she has discovered that accidents occur on a particular stretch of road at a rate of 5 per month. A lower speed limit is introduced on the road and Monisha wishes to investigate whether this has reduced the rate at which accidents occur. She records the number of accidents that occur in the next year.

  © Communication   MM Mathematical modelling   PS Problem solving

a Describe the test statistic, X.

b Write down a suitable null hypothesis.

c Write down a suitable alternative hypothesis.

d Is this a one-tailed or two-tailed test?

MM **5** Kash grows plants to sell in her nursery. She knows that the probability of a particular type of seed germinating is 0.8. Kash buys a new type of fertiliser and wishes to test whether it changes the probability of the seed germinating. She plants 100 seeds, using the new fertiliser, and records how many of them germinate.

a Describe the test statistic, X.

b Write down a suitable null hypothesis.

c Write down a suitable alternative hypothesis.

d Is this a one-tailed or two-tailed test?

MM **6** Approximately 10% of people are left-handed. Colin believes that people who study A-level maths are more likely to be left-handed. He decides to carry out a hypothesis test. At Colin's school, 50 people study A-level maths. He counts how many of these are left-handed.

a Describe the test statistic, X, and state its distribution, justifying your answer.

b State suitable null and alternative hypotheses, justifying your choices.

c State the distribution of X if your null hypothesis is true.

d Is this a one-tailed or a two-tailed test? Explain your answer.

MM **7** Over many years, it is found that during April, the probability that it will rain on a particular day is 0.35. Hans believes that it now rains less in April than it used to. He carries out a hypothesis test. Hans records the number of days in April on which it rains.

a Describe the test statistic, X, and state its distribution, justifying your answer.

b State suitable null and alternative hypotheses, justifying your choices.

c State the distribution of X if your null hypothesis is true.

d Is this a one-tailed or a two-tailed test? Explain your answer.

MM **8** Cars arrive at Hugo's fuel station at an average rate of 40 per hour. Road repairs are being carried out near the fuel station. Hugo decides to carry out a hypothesis test to decide if the number of cars arriving at his fuel station has been affected by the road repairs. He records the number of cars that arrive at his fuel station in an hour.

a Describe the test statistic, X, and state its distribution, justifying your answer.

b State suitable null and alternative hypotheses, justifying your choices.

c State the distribution of X if your null hypothesis is true.

d Is this a one-tailed or a two-tailed test? Explain your answer.

9 Nisha is a basketball player. When she shoots the ball, the probability of it going in the basket is 0.4. A new coach joins her team and tells her she should change her shooting technique. Nisha believes that the new technique has increased the probability of her shooting the ball into the basket.

a Nisha says she could use a binomial model to carry out a hypothesis test. Is she correct? Explain your answer.

b Give a reason why the Poisson distribution would not be suitable to use as a model for the hypothesis test.

c Nisha's coach says that he would use a two-tailed hypothesis test. Explain why he is wrong.

10 Read each of these statements, then say whether it is always true, sometimes true, or never true. Explain your answers.

a The null hypothesis states a theoretical probability.

b If you believe the value of the population parameter has increased, you should use a two-tailed test.

c When taken together, the null hypothesis and the alternative hypothesis cover all possible outcomes.

d To form null and alternative hypotheses, you need to know the probability distribution of the test statistic.

e The alternative hypothesis states that the population parameter is different from the value in the null hypothesis.

5.2 Significance levels and rejection regions

You need to decide if there is enough evidence from your experiment to reject the null hypothesis. If there is not enough evidence, the null hypothesis is accepted. Notice that you say you 'reject H_0', rather than 'accept H_1'.

> You can only accept or reject H_0 and state the evidence that supports this. As this is only an experiment from a sample of the population, you cannot categorically conclude from your findings.

Stop and think
Some statisticians prefer to say they 'do not reject H_0' rather than that they 'accept H_0'. Why do you think this is?

Significance levels

To decide if there is enough evidence to reject H_0, you must find out if the value of the test statistic you obtained from your experiment is unlikely. You use probability to make this decision.

A hypothesis test is given a significance level, α%. This is a chosen probability that implies an unlikely event. The significance level gives the probability that H_0 is rejected when it is true.

The most common significance levels are 1%, 5% and 10%. If your significance level is 1%, you would reject H_0 if your test result was in the most extreme 1% of possible outcomes.

A lower significance level means the test is more rigorous, but this can be a disadvantage. You are less likely to be able to reject H_0 and therefore your experiment may not conclude anything.

KEY INFORMATION

The term significance level, α, is used to refer to a chosen probability that implies an unlikely event. The significance level is the probability of rejecting H_0 when it is true.

The *p*-value of a test value

There are two ways of deciding whether your test result gives enough evidence to reject the null hypothesis. The first method is to calculate the probability of obtaining your test value, or more extreme values, when H_0 is true. This probability is known as the *p*-value.

For example, suppose you are carrying out a hypothesis test to decide if a die is biased towards 3. You decide to roll the die 12 times. In five of these rolls the die lands on 3.

The test statistic, *X*, is the number of times the die lands on 3 in 12 rolls.

If the null hypothesis is true, the probability of rolling a 3 is $\frac{1}{6}$.

The test statistic then has a binomial distribution, $X \sim B\left(12, \frac{1}{6}\right)$.

The test value is 5. Any values greater than 5 are considered to be more extreme than 5 as they are less likely. The *p*-value would be $P(X \geqslant 5)$.

If the *p*-value is less than the significance level, H_0 is rejected. If the *p*-value is greater than the significance level, H_0 is accepted.

In a two-tailed test, the significance level is split between the two tails of the distribution. This means the *p*-value must be compared to half the significance level, $\frac{\alpha}{2}$.

> **KEY INFORMATION**
>
> The term *p*-value is used to indicate a probability that you calculate after a given study. The *p*-value, or calculated probability, is the probability of finding the observed result, or more extreme values, when H_0 is true.

> It can be difficult to decide which values are 'more extreme' than your test value. It is helpful to think of them as being 'less likely' than your test value but in the same tail of the distribution.

> If the significance level was 5%, you would split this into 2.5% in the lower tail and 2.5% in the upper tail.

Carrying out a hypothesis test using the *p*-value

When carrying out a hypothesis test using the *p*-value, these steps should be clearly followed.

- Define the test statistic, *X*, and state its distribution.
- State H_0, H_1, whether the test is one-tailed or two-tailed, and the significance level.
- State the distribution of *X* if H_0 is true.
- State the criteria for rejection of H_0.
- Find the *p*-value of your test result and compare it to the significance level.
- Conclude if H_0 is accepted or rejected, referring to the context.

Example 3

To test whether his tetrahedral die is biased towards 1, Sunjit rolls it 8 times. 3 of these rolls the die lands on 1. Carry out a hypothesis test at the 5% significance level to test Sunjit's claim.

Solution

X = number of times die lands on 1 in 8 throws.

$X \sim B(8, p)$

Define the test statistic and its distribution.

$H_0: p = 0.25$

$H_1: p > 0.25$

State null and alternative hypotheses. See Example 1.

If you assume H_0 is correct, then $X \sim B(8, 0.25)$

State the distribution of the test statistic if the null hypothesis is true.

Use a one-tailed test at the 5% significance level.

State the type of test and the significance level.

The test value is $x = 3$, so you calculate $P(X \geq 3)$.

This is the probability of a result at least as extreme as 3.

H_0 will be rejected if $P(X \geq 3) < 0.05$.

State the criteria for rejection of H_0.

$P(X \geq 3) = 1 - P(X < 3) = 1 - [P(X = 0) + P(X = 1) + P(X = 2)]$

$P(X = 0) + P(X = 1) + P(X = 2)$

Calculate the required probability.

$= {}^8C_0(0.75)^8(0.25)^0 + {}^8C_1(0.75)^7(0.25)^1 + {}^8C_2(0.75)^6(0.25)^2$

$= 0.6785$

$1 - 0.6785 = 0.3215 = 0.322$ (3 s.f.)

$0.322 > 0.05$

Compare the p-value to the significance level. In this case it is greater, so H_0 is accepted.

H_0 is accepted. There is insufficient evidence at the 5% significance level that Sunjit's claim of the die being biased towards 1 is correct.

State whether the null hypothesis is accepted or rejected, relating your answer back to the original context.

Example 4

4% of components produced by a machine are faulty. Adjustments are made to the machine. 100 components are chosen at random and 2 are found to be faulty. Test, at the 5% significance level, if the adjustments have changed the percentage of components that are faulty.

Solution

The test statistic is the number of faulty components found out of 100.

$X \sim B(100, p)$

Define the test statistic and its distribution.

$H_0: p = 0.04$

$H_1: p \neq 0.04$

State null and alternative hypotheses. The null hypothesis is that the proportion of faulty components has not changed. The alternative hypothesis is that it has changed.

If H_0 is true, $X \sim B(100, 0.04)$.

State the distribution of the test statistic if the null hypothesis is true.

Use a two-tailed test at the 5% significance level.

State the type of test and the significance level.

The test value is $x = 2$, which is lower than expected, so look at the lower tail of the distribution and calculate $P(X \leq 2)$.

This is the probability of a result at least as extreme as 2.

H_0 is rejected if $P(X \leq 2) < 0.025$.

State the criteria for rejection of H_0.

$P(X \leq 2) = P(X = 0) + P(X = 1) + P(X = 2)$

$= {}^{100}C_0 (0.96)^{100}(0.04)^0 + {}^{100}C_1(0.96)^{99}(0.04)^1 + {}^{100}C_2(0.96)^{98}(0.04)^2$

$= 0.232$ (3 s.f.)

Calculate the required probability

$0.232 > 0.025$

Compare the p-value to the significance level. In this case it is greater.

H_0 is accepted. There is not sufficient evidence at the 5% level to suggest the adjustments to the machine have changed the percentage of components that are faulty.

State whether the null hypothesis is accepted or rejected, relating your answer back to the original context.

Rejection regions

The second method of deciding whether the null hypothesis can be rejected involves finding the rejection region (sometimes called the critical region). This involves calculating which values of the test statistic lead to H_0 being rejected.

To find the rejection region, p-values are calculated for different values of the test statistic. These are compared to the significance

KEY INFORMATION

The rejection region gives the values of the test statistic that lead to H_0 being rejected.

level of the test. A value of the test statistic lies in the rejection region if its p-value is less than the significance level.

In the upper tail of the distribution, the smallest value of the test statistic that lies in the rejection region is called the critical value. In the lower tail of the distribution, the critical value is the largest value that lies in the rejection region.

Carrying out a hypothesis test using the rejection region

When carrying out a hypothesis test using the rejection region, these steps should be clearly followed.

> Define the test statistic, X, and state its distribution.

> State H_0, H_1, whether the test is one-tailed or two-tailed, and the significance level.

> State the distribution of X if H_0 is true.

> State the criteria for rejection of H_0.

> Find the critical value of the test statistic. Define the rejection region and state whether your test value lies within it.

> Conclude if H_0 is accepted or rejected, referring to the context.

The values of X that lead to H_0 being accepted are known as the acceptance region.

Example 5

Hannah practises the guitar 3 days per week on average. She is preparing for an exam and decided to create a practice schedule with the aim of increasing the number of days per week on which she practises. She decides to test whether this has worked and carries out a hypothesis test at the 10% significance level.

a Set up Hannah's hypothesis test, and find the rejection region for the test.

b Hannah finds that, in a randomly chosen week, she practises on 5 days. Use your answer to part **a** to decide if there is evidence that her schedule has increased the number of days on which she practises.

Solution

a The test statistic is the number of days on which Hannah practises in a week $X \sim \text{Po}(\lambda)$.

$H_0: \lambda = 3$, $H_1: \lambda > 3$

If H_0 is true, $X \sim \text{Po}(3)$

Use a one-tailed test at the 10% significance level. H_0 is rejected if $P(X > x) < 0.1$.

> The null hypothesis is that Hannah still practises an average of three days per week. The alternative hypothesis is that she now practises an average of more than three days per week.

$P(X = 7) = e^{-3} \dfrac{3^7}{7!} = 0.0216$

$P(X = 6) = e^{-3} \dfrac{3^6}{6!} = 0.0504$

$P(X = 5) = e^{-3} \dfrac{3^5}{5!} = 0.1008$

$P(X \geqslant 6) = 0.0720$

$P(X \geqslant 5) = 0.1728$

$0.0720 < 0.1$ and $0.1728 > 0.1$

So the rejection region for the test is $X \geqslant 6$. •‒‒‒‒‒‒‒‒‒‒‒‒‒ $X = 6$ is the critical value.

b The test value is $x = 5$, which is not in the rejection region. Therefore H_0 is accepted and there is no evidence to suggest that the number of days on which Hannah practises is increased.

Exercise 5.2A

1 Karla catches the bus to school 5 days a week. Over a long period of time, Karla finds that the probability of her school bus being late in the morning is 0.2. A new bus company takes over the provision of her school bus. In the next week, her bus is late on 4 mornings. Karla tells the school that the new bus company is worse at arriving on time than the old one. Test Karla's claim at the 5% significance level.

2 Residents of a town who read the local newspaper complain that it has too many spelling mistakes in it. One resident records the number of spelling mistakes over a long period and finds there are an average of 4 mistakes per article. The editor of the paper makes changes to the proofreading process and claims the number of spelling mistakes has been reduced. In the next newspaper, the resident finds that there are an average of 2 mistakes per article. Test the editor's claim at the 1% significance level.

3 0.5% of the light bulbs produced by a factory are faulty. Changes are made to the manufacturing process and 1000 light bulbs are tested at random. One light bulb is found to be faulty. Test, at the 5% significance level, whether the manufacturing changes have made a difference to the proportion of faulty light bulbs that are produced.

4 When Marcus plays tennis, he serves an ace in 1 out of every 20 serves. In preparation for a tournament, he spends more time practising his serve. Marcus wants to see whether his serving has now improved. He decides to record the number of aces in the next 10 serves.

a Find the rejection region for a test at the 10% significance level.

b Marcus serves 2 aces in the next 10 serves. Use your answer to part **a** to decide if there is evidence that he has improved.

5 An emergency services call centre receives, on average, 5 calls per minute. The manager suspects that they receive fewer calls between 8 pm and 6 am.

a Find the rejection region for a test at the 5% significance level.

b The manager finds that an average of 3 calls per minute are received one night between 8 pm and 6 am. Use your answer to part **a** to test his claim.

6 Itsuko is an archer who hits the centre of the target 60% of the time. She is trialling a new bow to find out if it changes her accuracy. She fires 10 arrows with the new bow and records how many times she hits the centre of the target.

a Find the rejection region for a test at the 2% significance level.

b Itsuko hits the centre of the target 9 times. Use your answer to part **a** to decide if there is evidence that the new bow has changed her accuracy.

7 A delivery company promises that they deliver 95% of parcels the day after they are ordered. Jean-Luc orders 10 parcels from the delivery company and decides that if 8 or fewer of them are delivered the day after they are ordered, the company are not keeping their promise. Calculate the significance level of his test.

8 Aisha is worried about the speed of vehicles travelling through her village. She finds that, on average, 4 vehicles per hour are travelling through the village in excess of the speed limit. Aisha produces signs asking vehicles to slow down and wants to test whether these have worked. She decides that if 2 or fewer vehicles in an hour are travelling above the speed limit, her signs have worked. Calculate the significance level of her test.

9 Joel thinks a coin is biased towards heads. He throws the coin 10 times and records how many times it lands on heads. The rejection region for the test is chosen to be $X \geq 9$. Calculate the significance level of the test.

10 A random variable X has a Poisson distribution with mean λ. A single observation of X is taken and used to test the null hypothesis that $\lambda = 5$ against the alternative hypothesis that $\lambda < 5$. The rejection region for the test is chosen to be $X \leq 1$. Calculate the significance level of the test.

5.3 Using the normal approximation to the binomial or Poisson distribution

If the sample size for the hypothesis test is large, it can make the probability calculations very lengthy. In this case, the normal approximation to the binomial or Poisson distribution can be used. You must remember to apply a continuity correction as you are approximating a discrete distribution with a continuous distribution.

When carrying out hypothesis tests involving the normal distribution, it is often easiest to work out the critical values of the standardised test statistic Z. These can usually be found from the table of critical values of the normal distribution. They can then be used to define the rejection region.

> Refer to **Probability & Statistics 1, Chapter 4, The normal distribution** and **Chapter 1, The Poisson distribution**.

> **KEY INFORMATION**
>
> Remember if $X \sim B(n, p)$, $np \geq 5$ and $nq \geq 5$, then X can be approximated by $Y \sim N(np, npq)$.
>
> If $X \sim Po(\lambda)$ and λ is sufficiently large (greater than 15), then X can be approximated by $Y \sim N(\lambda, \lambda)$.

What other method could you use?

Example 6

In a driving test centre, the probability of passing your driving test first time is 54%. Abdel claims that his driving school is better than this. He records the first-time pass rate for 50 randomly selected learners he teaches. 31 of these pass.

Test Abdel's claim at the 5% level of significance.

Solution

The test statistic is the number of people, X, in a sample of 50 who pass first time. $X \sim B(50, p)$

H_0: $p = 0.54$, H_1: $p > 0.54$

If H_0 is true, $X \sim B(50, 0.54)$

Using the normal approximation:

$X \sim N(50 \times 0.54, 50 \times 0.54 \times 0.46)$

$X \sim N(27, 12.42)$

The standardised test statistic is $Z \sim N(0, 1)$

Use a one-tailed test at the 5% significance level.

$P(Z \le z) = 0.95 \therefore z = 1.645$

H_0 is rejected if $Z > 1.645$.

$z = \dfrac{30.5 - 27}{\sqrt{12.42}} = 0.993$

$0.993 < 1.645$ so does not lie within the rejection region.

H_0 is accepted. There is no evidence at the 5% significance level that Abdel's driving school has a greater proportion of passes than the driving test centre does.

> The null hypothesis is that the probability of passing first time is 0.54. The alternative hypothesis is that it is greater than 0.54.

> You need to consider the upper tail, to correspond with H_1. $P(Z > 1.645) = 0.05$ so values of Z greater than 1.645 lie in the rejection region.

> A continuity correction is applied here.

Example 7

Over a long period, it is found that accidents occur at a highway intersection at the rate of 20 per year. New signage is put up and the highway agency believes this has changed the yearly rate of accidents. In the next six months, 7 accidents occur at the intersection. Test, at the 5% significance level, whether this supports the highway agency's belief.

Solution

The test statistic is the yearly rate of accidents. $X \sim \text{Po}(\lambda)$

H_0: $\lambda = 20$, H_1: $\lambda \neq 20$

If H_0 is true, $X \sim \text{Po}(20)$

Using the normal approximation:

$X \sim N(20, 20)$

The standardised test statistic is $Z \sim N(0, 1)$

Use a two-tailed test at the 5% significance level.

$P(Z \leqslant z) = 0.975 \Rightarrow z = 1.96$

H_0 is rejected if $Z < -1.96$.

$z = \dfrac{14.5 - 20}{\sqrt{20}} = -1.230$

$-1.230 > -1.96$ so does not lie within the rejection region.

H_0 is accepted. There is no evidence at the 5% significance level that the signs have changed the yearly rate of accidents.

> The null hypothesis is that the yearly rate of accidents is still 20. The alternative hypothesis is that it is not 20.

> The lower tail is looked at as the test value is smaller than the expected value. $P(Z < -1.96) = 0.025$ so values of Z less than -1.96 lie in the rejection region.

> $X = 14$ is used as the test value as the statistic is the yearly rate of accidents, then a continuity correction is applied.

Exercise 5.3A

1 Mario owns a restaurant. Over a long period of time, he has found that 20% of customers order a vegetarian meal. Mario decides to offer a free drink with every vegetarian meal sold. 35 of the next 100 customers order a vegetarian meal. Mario believes that the free drink offer has increased the number of vegetarian meals sold. Test his claim at the 5% significance level.

2 Mai runs a news stand. On average, she sells 30 copies of *The Town Observer* newspaper per day. The publishers of *The Town Observer* raise the price of the newspaper and Mai believes that she is now selling fewer copies. On a particular day, she sells 24 copies of *The Town Observer*. Test Mai's claim at the 1% significance level.

3 A pet food manufacturer claims that exactly 85% of cats prefer their brand of food, *Active Cat*, to the brand produced by their rival. Saskia runs a cattery and believes that the manufacturer's claim is incorrect. She offers both brands of food to a random sample of 50 cats and finds that 38 cats preferred the *Active Cat* brand. Using a test at the 5% significance level, decide whether there is evidence that Saskia is right to think the manufacturer's claim is incorrect.

4   Caitlin is working on an alarm for her Electronics project. It is rather unreliable and the probability that it works in a particular test is 0.4. Caitlin makes changes to her circuit and carries out 20 tests. She decides that the changes have improved the reliability of her alarm. Find the rejection region for a hypothesis test at the 5% significance level.

5   Imran works in a call centre. He currently takes an average of 50 calls per day. Imran is hoping to increase the average number of calls he takes in a day so he can earn a bonus. Imran records the number of calls he takes in the first 5 days in April and carries out a hypothesis test at the 10% significance level to find out if he is on course for a bonus. Find the rejection region for the test.

6   A delivery company owns a large number of vans. Over a long period of time, it is found that the vans break down at the rate of 21 per month. In a randomly chosen month, there are 27 breakdowns and the company believes that the vans are now breaking down more often. Calculate the significance level of this test.

7   Neave thinks a tetrahedral die is biased against 3. She rolls the die 60 times. It lands on 3 in 10 of these rolls. Neave says this means her die is biased against 3. Calculate the significance level of this test.

8   A company hires out bicycles. Over a long period of time, it is found that the company hires out an average of 25 bicycles per day. In a particular week, the company hires out 210 bicycles. They say this means they have increased the number of bicycles they hire out. Calculate the significance level of this test.

9   A teacher reads an article which claims that 20% of school students regularly read a newspaper. She believes that less than 20% of the pupils in her school regularly read a newspaper. She takes a random sample of 30 students and records how many regularly read a newspaper. Find the rejection region for a hypothesis test at the 5% significance level.

10 A real estate agent sells, on average, 40 properties per month. He believes that a new marketing campaign has increased the number of properties he sells and records the number of properties he sells in the next month. Find the rejection region for a hypothesis test at the 2% significance level.

Mathematics in life and work: Group discussion

You work for a highways agency and have been given the task of investigating whether a new road has improved traffic conditions in a town. You have access to data that was collected before the road was built. You decide to focus your first investigation on the number of traffic accidents per month in the town.

1   What probability distribution would best model this, and why?

2   How long would you want to collect data for before making your comparison?

3   What might your null and alternative hypotheses be?

4   Do you think a one-tailed or two-tailed test would be more appropriate? Why?

5   What significance level might you wish to use for your test? Why?

6   Who might be interested in your results?

5.4 The population mean

In **Chapter 4, Sampling and estimation**, you learnt that the sample mean has a normal distribution if:

> the population has a normal distribution

> in cases where the population does not have a normal distribution, the sample size is sufficiently large (the Central Limit Theorem).

You also learnt how to calculate probabilities of the sample mean, taking particular values by using normal distribution tables. You can apply this to hypothesis tests concerning the population mean. As you are using the normal distribution, finding the rejection region using critical values of Z is the easiest method.

> **KEY INFORMATION**
>
> $$\bar{X} \sim N\left(\mu, \frac{\sigma^2}{n}\right)$$

Stop and think What other method could you use?

Example 8

The masses of adult male silverback gorillas in the wild are known to be normally distributed with mean 230 kg and standard deviation 21.3 kg. A biologist suspects that the mass of adult male silverback gorillas is decreasing. A random sample of 14 males are weighed and their mean mass is found to be 218 kg. Does this provide evidence, at the 5% significance level, that the biologist's suspicion is correct?

Solution

The test statistic is the sample mean $\bar{X} \sim N\left(\mu, \frac{21.3^2}{14}\right)$

$H_0: \bar{X} = 230$, $H_1: \bar{X} < 230$

If H_0 is true, $\bar{X} \sim N\left(230, \frac{21.3^2}{14}\right)$.

The standardised test statistic is $Z \sim N(0, 1)$.
Use a one-tailed test at the 5% significance level.

$P(Z \leqslant z) = 0.95 \Rightarrow z = 1.645$

H_0 is rejected if $Z < -1.645$.

$z = \dfrac{218 - 230}{\left(\frac{21.3}{\sqrt{14}}\right)} = -2.108$

$-2.108 < -1.645$ so lies in the rejection region.

H_0 is rejected; there is evidence at the 5% significance level that the mean mass of adult male silverback gorillas has decreased.

The null hypothesis is that the mean mass is still 230 kg. The alternative hypothesis is that it is less than 230 kg.

You are looking at the lower tail of the distribution to correspond with H_1. $P(Z < -1.645 = 0.05)$, so values of Z less than -1.645 lie in the rejection region.

Example 9

The time taken for a bus to go from the centre of Riyadh to King Khalid airport has a mean of 28 minutes at rush hour, with standard deviation 7 minutes. A new road network has been created which is intended to speed the journey up for rush-hour traffic. A group of students monitor the traffic flow and claim that the journey is now taking longer than 28 minutes. The students take 250 observations at rush hour and find the mean time taken is 29.3 minutes.

a Test the students' claim at the 1% level of significance.

b State, with a reason, whether it was necessary to use the Central Limit Theorem in part **a**.

Solution

a The test statistic is the sample mean $\bar{X} \sim N\left(\mu, \dfrac{7^2}{250}\right)$

$H_0: \bar{X} = 28$, $H_1: \bar{X} > 28$

> The null hypothesis is that the mean journey time is equal to 28 minutes. The alternative hypothesis is that it is greater than 28 minutes.

If H_0 is true, $\bar{X} \sim N\left(28, \dfrac{7^2}{250}\right)$.

The standardised test statistic is $Z \sim N(0, 1)$.

Use a one-tailed test at the 1% significance level.

$P(Z \leqslant z) = 0.99 \Rightarrow z = 2.326$

H_0 is rejected if $Z > 2.326$.

> You are looking at the upper tail of the distribution, to correspond with H_1. $P(Z > 2.326) = 0.01$, so values of Z greater than 2.326 lie in the rejection region.

$z = \dfrac{29.3 - 28}{\left(\dfrac{7}{\sqrt{250}}\right)} = 2.936$

$2.936 > 2.326$ so z lies in the rejection region.

H_0 is rejected; there is evidence at the 1% significance level that the mean length of bus journeys is now greater than 28 minutes.

b It was necessary to use the Central Limit Theorem in part **a**, as you do not know if the population is normally distributed.

Exercise 5.4A

 1 A machine is designed to make pins with a mean mass of 2 g and a standard deviation of 0.05 g. The distribution of the masses of the pins is a normal distribution.

During an experiment, a student weighs a random sample of 30 pins and finds their total mass to be 59.4 g. Conduct a hypothesis test at the 5% significance level to find whether this provides evidence of a decrease in the mean mass of the pins.

MM **2** A certain breed of guinea pig has a mass that is normally distributed with a mean of 750 g and a standard deviation of 11.2 g. A breeder is not convinced her guinea pigs are of this breed type. In order to test this, she collects a sample of 8 and weighs them, with the following results.

752, 738, 756, 771, 729, 756, 739, 749

Carry out a hypothesis test at the 10% significance level and write down your conclusion.

MM **3** A machine at a store puts paint into cans. The store owner electronically controls the mean volume of the cans. The standard deviation is fixed at 0.11 litres. The mean volume is suspected to be lower than it should be. A random sample of 42 cans has a total volume of 333.9 litres. Is there evidence at the 5% level to support the suspicion if the mean volume is set to 8 litres?

MM **4** A local theatre group needs to buy some new stage lighting. The supplier of Brite light bulbs claims that the mean life of their light bulbs is 288 hours. The theatre group buys 150 of their bulbs but finds the mean only to be 254.5 hours, with a sample standard deviation of 11 hours. Is there evidence at the 1% level that the mean is lower than 288 hours?

MM **5** In order for a local shop to sell samphire to a speciality restaurant it must be of a certain length. The requirements of the lengths are that they should be normally distributed with mean 16 cm and standard deviation 0.8 cm. The shop manager finds a new area to collect from and wants to know if the samphire would be suitable to sell to the restaurant. She collects and measures 80 samples and finds that the total of their lengths is 1256 cm. Carry out a test at the 5% level. What should the shop manager conclude?

MM **6** The heights of marigold plants grown by Kash in her nursery are known to be normally distributed with mean 31 cm. Kash tries a new fertiliser on her plants and then measures the heights, in cm, of a random sample of 10 mature plants and records them as follows.

30.9, 35.2, 32.5, 28.4, 30.5, 33.4, 31.5, 29.7, 32.8, 34.6

   a Calculate unbiased estimates of the population mean and variance.

   b Use your answers to part **a** to test at the 2% significance level whether the new fertiliser has changed the mean height of the plants.

C **7** Andy goes to work on 5 days each week. Over a long period of time Andy finds that the mean length of his journey between home and work, in both directions, is 41 minutes. He decides to try a new route in the hope of reducing his travelling time.

He records the length of each journey every day for 4 weeks. His journey times are given by the following summary statistics.

$$\sum x = 1520 \quad \sum x^2 = 58\,760$$

   a Test, at the 1% significance level, whether Andy's new route has reduced his journey time to work.

   b State, with reasons, whether you used the Central Limit Theorem in part **a**.

PS **8** Fabric should be sold in 1.5 metre lengths. The lengths are normally distributed. It is believed that the mean length of a piece of fabric is more than 1.5 metres. A random sample of 5 lengths is taken and the following summary statistics are obtained.

$$\sum x = 7.62, \quad \sum x^2 = 11.65$$

    a The critical value is given as $z = 1.960$. What is the significance level of the test?

    b Find the value of the standardised test statistic, z.

    c What should the test conclude? Explain your answer.

(PS) **9** A random variable X has an unknown distribution. The standard deviation of X is known to be 2. It is believed that the mean of X is 10. A random sample of 30 observations of X is taken. The mean of this sample is 10.72. The conclusion is that the mean of X is greater than 10.

    a Calculate the significance level of this test.

    b Explain how the Central Limit Theorem was used in this test.

(MM) **10** Fen makes jam to sell. The amount of jam in a jar is normally distributed and should have a mean mass of 450 g. Fen wishes to test if her jars contain the correct amount of jam. She weighs the jam in a random sample of 10 jars with the following results.

    451   452   450   449   450   451   453   452   448   452

    a Use the sample data to calculate unbiased estimates of the population mean and standard deviation.

    b Test, at the 5% significance level, whether the mean mass of jam in a jar is 450 g.

    c Explain whether you needed to use the Central Limit Theorem in your calculations.

5.5 Type I and Type II errors

When a hypothesis test is carried out, there are four possible outcomes:

> H_0 is true and the test leads you to accept H_0.

> H_0 is true and the test leads you to reject H_0.

> H_0 is false and the test leads you to accept H_0.

> H_0 is false and the test leads you to reject H_0.

The first and fourth outcomes lead to a correct conclusion, but the second and third lead to an incorrect conclusion: in other words, an error.

A Type I error is made if H_0 is rejected when it is true. You already know that the significance level tells you the probability of rejecting H_0 when it is true. So the probability of a Type I error is the same as the significance level of the test.

A Type II error is made if H_0 is accepted when it is false. You can also think of this as H_0 being accepted when H_1 is true. This means that you can only calculate the probability of a Type II error if a specific value is stated for H_1. The method is the same as for calculating the p-value, except the value of the parameter in H_1 is used rather than the value in H_0.

It is not possible to make a Type I error if H_0 is accepted.

It is not possible to make a Type II error if H_0 is rejected.

KEY INFORMATION

P(Type I error) = P(H_0 rejected when it is true) = α

P(Type II error) = P(H_0 accepted when it is false) = P(H_0 accepted given that H_1 is true)

Example 10

Khara has a tetrahedral die. She suspects it is biased. She rolls the die 20 times and records how many times it lands on 1.

a With reference to this context, explain how a Type I error could be made.

b With reference to this context, explain how a Type II error could be made.

Solution

a A Type I error occurs if the null hypothesis is rejected when it is true. In this context, a Type I error could be made if Khara's test concludes that the probability of the die landing on 1 is not 0.25 when in fact it is 0.25.

b A Type II error occurs if the null hypothesis is accepted when it is false. In this context, a Type II error could be made if Khara's test concludes that the probability of the die landing on 1 is 0.25 when in fact it is not 0.25.

Example 11

Pat is trying to find out if a coin is biased. She throws the coin 10 times and it lands on tails 8 times. Pat says this means the coin is biased towards tails.

a Calculate the significance level of Pat's test.

b Hence write down the probability of a Type I error and explain what it means in this context.

c The coin is in fact biased so the probability of it landing on tails is 0.75. What is the probability that Pat makes a Type II error?

d Explain what a Type II error means, referring to the context of this test.

Solution

a This is a one-tailed test as Pat says the probability of tails is greater than 0.5. The rejection region for the test is $X \geqslant 8$. So the significance level of the test is $P(X \geqslant 8)$.

$P(X \geqslant 8) = P(X = 8) + P(X = 9) + P(X = 10)$

$= {}^{10}C_8(0.5)^2 (0.5)^8 + {}^{10}C_9(0.5)^1 (0.5)^9$

$+ {}^{10}C_{10}(0.5)^0 (0.5)^{10}$

$= 0.0547$ (3 s.f.)

The significance level is 5.47%.

b The probability of a Type I error is the same as the significance level of the test. So the probability of a Type I error is 0.0547.

A Type I error occurs if the null hypothesis is rejected when it is true. In this context a Type I error would occur if Pat concludes that the coin is biased towards tails when in fact it is fair.

c The probability of a Type II error is $P(X < 8)$ given that the probability of tails is 0.75.

$$P(X \geqslant 8) = P(X = 8) + P(X = 9) + P(X = 10)$$
$$= {}^{10}C_8(0.25)^2 (0.75)^8 + {}^{10}C_9(0.25)^1 (0.75)^9$$
$$+ {}^{10}C_{10}(0.25)^0 (0.75)^{10}$$
$$= 0.526 \text{ (3 s.f.)}$$

$$P(X < 8) = 1 - P(X \geqslant 8)$$
$$= 1 - 0.526$$
$$= 0.474$$

d A Type II error occurs if the null hypothesis is accepted when it is false. In this test, a Type II error would occur if Pat concluded that the coin is fair when in fact it is biased.

Example 12

Max is a reprographics technician. Over a long period of time, the photocopier he uses experiences a paper jam an average of 8 times per day. He orders a different brand of paper in the hope of reducing the number of paper jams per day. Max decides that if the number of jams in a particular day is 3 or fewer, the new paper has reduced the number of jams.

a Find the probability of making a Type I error, and hence find the significance level of the test.

b If in fact, the number of jams is reduced to an average of 6 per day, find the probability of making a Type II error.

Solution

a The test statistic is the number of jams in one day:
$X \sim \text{Po}(\lambda)$

$H_0: \lambda = 8$, $H_1: \lambda < 8$

If H_0 is true, $X \sim \text{Po}(8)$

Use a one-tailed test at the $\alpha\%$ significance level. H_0 is rejected if $X \leqslant 3$.

> The null hypothesis is that the number of jams is still 8 per day. The alternative hypothesis is that it is less than 8.

$$P(X=0) = e^{-8}\frac{8^0}{0!} = 0.0003$$

$$P(X=1) = e^{-8}\frac{8^1}{1!} = 0.0027$$

$$P(X=2) = e^{-8}\frac{8^2}{2!} = 0.0107$$

$$P(X=3) = e^{-8}\frac{8^3}{3!} = 0.0286$$

$$P(X \leqslant 3) = 0.0424$$

The probability of a Type I error is 0.0424 and the significance level is $\alpha = 4.24\%$.

> The significance level is the probability of a Type I error.

b The probability of a Type II error is $P(X > 3)$ if $X \sim \text{Po}(6)$.

> This is the probability that H_0 is accepted when it is false.

$$P(X > 3) = 1 - P(X \leqslant 3)$$

$$P(X=0) = e^{-6}\frac{6^0}{0!} = 0.0025$$

$$P(X=1) = e^{-6}\frac{6^1}{1!} = 0.0149$$

$$P(X=2) = e^{-6}\frac{6^2}{2!} = 0.0446$$

$$P(X=3) = e^{-6}\frac{6^3}{3!} = 0.0892$$

$$P(X \leqslant 3) = 0.1512$$

$$P(X > 3) = 1 - 0.1512 = 0.8488$$

The probability of a Type II error is 0.849.

Exercise 5.5A

1 At a medical clinic, the probability that a patient does keep their appointment is 0.15. The booking system is changed to try to reduce the number of patients who do not turn up. A hypothesis test is carried out at the 1% significance level to see if the changes have worked.

a Write down the probability that a Type I error is made in the test.

b Explain what a Type I error would mean in this context.

c Explain what a Type II error would mean in this context.

2 Carpet is produced at a factory. On average there are 2 flaws per metre of carpet. A new quality control procedure is introduced to reduce the amount of flaws in the carpet. A hypothesis test is carried out at the 2% significance level to see if the changes have worked.

a Write down the probability that a Type I error is made in the test.

b Explain what a Type I error would mean in this context.

c Explain what a Type II error would mean in this context.

3 A train is late 20% of the time. A new operator takes over the running of the train and claims the train is now on time more often. A hypothesis test is carried out to see if the operator is correct.

   a The probability of a Type I error is found to be 0.042. Write down the significance level of the test.

   b Explain what a Type I error would mean in this context.

   c Explain what a Type II error would mean in this context.

4 The elevators in an office building break down on average 3 times a week. A new maintenance schedule is introduced and it is believed that the elevators now break down less often. A hypothesis test is carried out to investigate if this is true.

   a The probability of a Type I error is found to be 0.0674. Write down the significance level of the test.

   b Explain what a Type I error would mean in this context.

   c Explain what a Type II error would mean in this context.

5 In a factory that manufactures scissors, the probability that a pair of scissors is faulty is 0.05. After changes to the manufacturing process, a random sample of 100 pairs of scissors is checked and 2 are found to be faulty.

   a A hypothesis test at the 10% significance level is carried out to see if the adjustments have reduced the proportion of faulty pairs of scissors that are produced. Write down the probability of a Type I error.

   b The quality control manager decides that the results from the sample show that the proportion of scissors that are faulty has reduced. Following the adjustments, the proportion of pairs of scissors that are faulty is 3%. Calculate the probability of a Type II error.

6 Large trucks pass through a village at an average rate of 4 per hour. Marcio runs a campaign to reduce the number of large trucks passing through the village and decides it has been successful if fewer than 48 lorries pass through the village in a 24-hour period. Calculate the probability of a Type I error and state the significance level of the test.

7 A chocolate company runs a promotion where 5% of bars of chocolate contain a voucher for a free bar of chocolate. Elena buys 20 bars of chocolate and finds two that have a voucher inside. Elena says that this means that more than 5% of bars of chocolate contain a voucher. Calculate the probability of a Type I error and state the significance level of the test.

8 Yoshi suspects a certain coin is biased and that p, the probability of it landing on tails, is greater than 0.5. He spins the coin 200 times and records that it lands on tails 120 times. Yoshi then carries out a hypothesis test at the 6% significance level.

   a Carry out Yoshi's test.

   b Explain whether a Type I or Type II error could have been made, referring to the context of the test.

9 Rami owns an organic farm. He finds that the average number of aphids on his cabbage plants is 250. Rami buys some ladybugs to try to control the aphids. After a month he counts the number of aphids on one of his plants and finds that there are 210. Ravi decides that this means that the ladybugs have reduced the number of aphids.

    a Calculate the probability of a Type I error and explain what it means in this context. Hence state the significance level of Rami's test.

    b If in fact the ladybugs have reduced the average number of aphids on a plant to 200, calculate the probability of a Type II error.

 10 A survey conducted 10 years ago showed that the mean number of cars owned by households in a town was 1.34 with standard deviation 1.05. A random sample of 50 households in the town are surveyed and it is found that the mean number of cars owned by these households is now 1.74.

    a Test at the 4% significance level whether there is evidence that the mean number of cars per household has increased.

    b Explain whether it is possible that a Type I error was made when carrying out the test.

    c Explain what would be meant by a Type II error in this context, and give the range of values of the standardised test statistic that could result in a Type II error.

Mathematics in life and work: Group discussion

You work for a highways agency and have been given the task of investigating whether a new road that avoids the town centre has improved traffic conditions in a town. You have access to data that was collected before the new road was built. You have now decided to investigate whether the new road has reduced the mean mass of vehicles passing through the town.

1 What statistics would you need in order to carry out this investigation?

2 What would be the test statistic?

3 What might your null and alternative hypotheses be?

4 Under what circumstances would you need to use the Central Limit Theorem?

5 What would Type I and Type II errors mean in this context?

6 Under what circumstances could these errors be made?

SUMMARY OF KEY POINTS

› A statistical hypothesis is a statement about the value of a population parameter.

› A hypothesis test uses sample data to decide whether there is evidence to support the hypothesis. The statistic calculated from the sample data is known as the test statistic.

› The null hypothesis, H_0, states the expected or theoretical value of a population parameter. The alternative hypothesis, H_1, states that the parameter has changed or is different from the expected value.

› In a one-tailed test the alternative hypothesis states that the parameter is either greater than or less than the value given in the null hypothesis. In a two-tailed test, the alternative hypothesis states only that the parameter is not equal to the value given in the null hypothesis.

› The significance level of a hypothesis test gives the probability that H_0 is rejected when it is true.

› The p-value of a test value gives the probability of obtaining that value or more extreme values, if H_0 is true.

› If the p-value is less than the significance level, H_0 is rejected. If it is greater than the significance level, H_0 is accepted. In a two-tailed test, the p-value is compared to half of the significance level.

› The rejection region (or critical region) comprises the values of the test statistic that will lead to H_0 being rejected. The acceptance region comprises the values of the test statistic that will lead to H_0 being accepted.

› The critical values of the test statistic are the largest value that falls into the rejection region in the lower tail and the smallest value that falls into the rejection region in the upper tail.

› A Type I error occurs if H_0 is rejected when it is true. The probability of a Type I error is given by the significance level.

› A Type II error occurs if H_0 is accepted when it is false. The probability of a Type II error is calculated in the same way as the p-value, but assuming that H_1 is true rather than H_0.

EXAM-STYLE QUESTIONS

(C) 1 $X \sim B(20, p)$. A random observation, x, of X is taken. Given that $x = 18$, test at the 4% significance level the null hypothesis that $p = 0.8$ against the alternative hypothesis that $p > 0.8$.

(C) 2 $X \sim Po(\lambda)$. A random observation, x, of X is taken. Find the rejection region for a test at the 10% significance level of the null hypothesis that $\lambda = 5$ against the alternative hypothesis that $\lambda < 5$.

(C) 3 $X \sim Po(\lambda)$. A random observation, x, of X is taken. If $x = 2$, test at the 6% significance level the null hypothesis that $\lambda = 3$ against the alternative hypothesis that $\lambda \neq 3$.

(C) 4 $X \sim B(50, p)$. A random observation, x, of X is taken. Use a suitable approximation to find the rejection region for a test at the 5% significance level of the null hypothesis that $p = 0.4$ against the alternative hypothesis that $p < 0.4$.

(C) 5 $X \sim N(\mu, 1.3^2)$. A random sample of 20 observations of X is taken, and it is found that $\bar{x} = 15.2$. Test, at the 8% significance level, the null hypothesis that $\mu = 12$ against the alternative hypothesis that $\mu > 12$.

(MM) 6 Davide runs a market stall selling fruit and vegetables. He buys tomatoes in boxes from the wholesaler. Over a long period of time, he finds that an average of 4 tomatoes per box are damaged. Davide complains to the wholesaler who promises to reduce the number of damaged tomatoes they sell. In the next 10 boxes he buys, a total of 25 tomatoes are damaged. Test, at the 5% significance level, whether there is evidence that the wholesaler has kept their promise.

(MM) 7 Alex is managing a political campaign. Previous surveys have shown that 65% of people support his candidate. Alex has organised a delivery of leaflets with information about his candidate to a particular area. He wishes to know if the leaflets have increased support and decides to ask 10 random people in the area if they support his candidate.

  a Find the rejection region for a test at the 2% significance level.

  b Eight of the people Alex asks support his candidate. Is there evidence that the leaflets have increased support?

(MM) 8 Following a change in schedules, a delivery driver wishes to test whether the mean distance he drives in a week has changed. He notes the distance, x km, that he drives in 30 randomly chosen weeks. The results are noted as follows.

$$\sum x = 13\,526, \quad \sum x^2 = 9\,175\,354$$

  a Calculate unbiased estimates of the population mean and variance.

  b In the past the mean distance he drove in a week was 650 km. Carry out a test at the 5% significance level to find out if the mean distance he drives each week has changed.

  c Explain whether you needed to use the Central Limit Theorem in part **b**.

(MM) 9 10% of the biscuits produced in a factory are damaged. Changes are made to the manufacturing process to reduce the proportion of biscuits that are damaged. A random sample of 1000 biscuits is taken and it is found that 80 are damaged. Test, at the 2% significance level, if the changes have reduced the proportion of biscuits that are damaged.

(MM) 10 Water in a lake contains on average 60 bacteria per litre. A biologist tests a random litre of the water and finds that there are 75 bacteria in this sample. She says this means the number of bacteria has increased. Test this claim at the 5% significance level.

(MM) 11 A call centre receives an average of 30 calls per hour. The manager believes they receive fewer calls at night. One night, an hour is selected at random and the number of calls recorded.

  a Find the rejection region for a hypothesis test at the 5% significance level.

  b 20 calls are received in that hour. Do you agree with the manager's claim? Explain your answer.

  c Explain whether a Type I or Type II error could have been made, referring to the context of the test.

12 In the last population census, 5% of the population of a town said that they usually travelled to work by bicycle. The town council introduces a 'cycle to work' scheme to try to increase the proportion of people who travel to work by bicycle. A random sample of 500 people is taken and 35 say that they usually travel to work by bicycle. The council says that this means the proportion of people who usually travel to work by bicycle has increased.

   a Calculate the significance level of the test and hence write down the probability of a Type I error.

   b In fact the proportion of people who usually travel to work by bicycle is now 8%. Calculate the probability of a Type II error.

13 Meera is a javelin thrower. To qualify for a competition she must throw over 25 metres. The proportion of Meera's javelin throws that are over 25 metres is 0.6. She trains for an extra hour per day for a month. She then decides that if she throws the javelin over 25 metres in at least 8 of the next 10 throws, she has improved.

   a Calculate the significance level of the test. Hence write down the probability of a Type I error.

   b The proportion of Meera's throws that are over 25 metres is in fact now 0.75. Calculate the probability of a Type II error and explain what it means in this context.

14 Kimi runs a second-hand bookstore. His weekly profit is normally distributed with mean $500 and standard deviation $60. Kimi decides to change the opening hours and, in the next 10 weeks, makes a total profit of $5500.

   a Test, at the 5% significance level, whether the new opening hours have changed Kimi's mean weekly profit.

   b Write down the probability that a Type I error was made in part **a**.

   c Given that Kimi's mean weekly profit has in fact increased to $540, calculate the probability that a Type II error was made in part **a**.

15 Jules works for an average of 35 hours per week. Following a restructure in his department, his role is changed but he is paid the same as before. In the next 20 weeks, Jules works a total of 775 hours. Jules is considering looking for another job as he believes he is now working for more than 35 hours per week for the same pay.

   a Test Jules' claim at the 10% significance level.

   b Jules is now working an average of 38 hours per week. Calculate the probability that a Type II error was made in part **a**.

16 The masses of potatoes grown on a farm are normally distributed with mean 138 grams and standard deviation 21 grams. Nutrients are added to the soil and a random sample of n potatoes taken, which are found to have mean mass 141.8 grams. A hypothesis test is then carried out to find whether the nutrients have increased the mean mass of the potatoes.

   a The critical value of the test statistic, Z, is 1.645 to 3 d.p. State the significance level of the test.

   b Given that the test value of Z is 1.13, find the value of n.

   c What assumption did you need to make in part **a**?

   d Carry out the hypothesis test.

MM **17** Over many years, the mean mark achieved by all students who sit an examination is 67 with standard deviation 7. It is thought this year's exam is easier than usual. To test this hypothesis, a random sample of 200 students is chosen and their mean mark found to be 72.

    a The critical value of the test statistic, Z, is 2.326 to 3 d.p. State the significance level of the test and the probability of a Type I error.

    b Carry out the test and explain whether there is evidence that this year's exam is easier than usual.

    c State, with a reason, whether you needed to use the Central Limit Theorem in parts **a** and **b**.

18 The mean weekly profit made by a hotel is $2000, with standard deviation $300. The hotel owner decides to pay for membership of a popular holiday booking website. After joining the website, the mean weekly profit in a random sample of 40 weeks is $3000.

    a Stating an assumption you have made, test at the 1% significance level whether the mean weekly profit has increased after joining the website.

    b State, with a reason, whether you needed to use the Central Limit Theorem in part **a**.

    c Explain whether or not it is possible that a Type I error was made in part **a**.

MM **19** It is known that 60% of people living in a town support the building of a new development of houses. Following a campaign by an environmental group, the town council believe that the proportion of people supporting the development has changed. The council asks a random selection of 10 people from the town and carries out a test at the 5% significance level.

    a Find the rejection region for the test.

    b Three of the 10 people say they support the development. State which error, Type I or Type II, might be made, explaining your answer.

Mathematics in life and work

You have been asked to investigate whether the building of a new road has affected the profits of businesses in a town. Before the road was built, the mean weekly profit of businesses in the town was $2500, with standard deviation $750. In a random sample of 50 weeks following the road being completed, the mean weekly profit of the businesses in the town was $2750.

1 Carry out a hypothesis test at the 5% significance level to investigate whether there is evidence that the mean weekly profit has changed.

2 Explain whether it was possible that a Type I error was made in part **a**.

3 Explain whether the Central Limit Theorem was used in part **a**.

4 The mean weekly profit has in fact increased to $2550 since the road was built. Calculate the probability that a Type II error was made in part **a**.

SUMMARY REVIEW

Practise the key concepts and apply the skills and knowledge that you have learned in the book with these carefully selected past paper questions supplemented with exam-style questions and extension questions written by the authors.

Warm-up Questions	A Level Questions	Extension Questions

Three Cambridge A Level Mathematics past paper questions based on prerequisite skills and concepts that are relevant to the main content of this book.

Selected past paper exam questions and exam-style questions on the topics covered in this syllabus component.

Extension questions that give you the opportunity to challenge yourself and prepare you for more advanced study.

Warm-up Questions

1 A team of 4 is to be randomly chosen from 3 boys and 5 girls. The random variable X is the number of girls in the team.

    i Draw up a probability distribution table for X.   **[4]**

    ii Given that $E(X) = \frac{5}{2}$, calculate $Var(X)$.   **[2]**

Cambridge International AS & A Level Mathematics 9709 Paper 61 Q3 Nov 2011

2 It is given that $X \sim N(30, 49)$, $Y \sim N(30, 16)$ and $Z \sim N(50, 16)$. On a single diagram, with the horizontal axis going from 0 to 70, sketch three curves to represent the distributions of X, Y and Z.   **[3]**

Cambridge International AS & A Level Mathematics 9709 Paper 61 Q1 Nov 2013

3 The random variable X has the distribution $N(\mu, \sigma^2)$. It is given that $P(X < 54.1) = 0.5$ and $P(X > 50.9) = 0.8665$. Find the values of μ and σ.   **[4]**

Cambridge International AS & A Level Mathematics 9709 Paper 61 Q2 Nov 2015

A Level Questions

 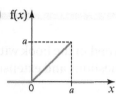

The random variable X has probability density function, f(x), as shown in the diagram, where a is a constant. Find the value of a and hence show that E(X) = 0.943 correct to 3 significant figures.   [5]

Cambridge International AS & A Level Mathematics 9709 Paper 71 Q1 June 2015

2 Failures of two computers occur at random and independently. On average the first computer fails 1.2 times per year and the second computer fails 2.3 times per year. Find the probability that the total number of failures by the two computers in a 6-month period is more than 1 and less than 4.   [4]

Cambridge International AS & A Level Mathematics 9709 Paper 71 Q1 Nov 2015

3 The probability that a randomly chosen plant of a certain kind has a particular defect is 0.01. A random sample of 150 plants is taken.

i Use an appropriate approximating distribution to find the probability that at least 1 plant has the defect. Justify your approximating distribution.   [4]

The probability that a randomly chosen plant of another kind has the defect is 0.02. A random sample of 100 of these plants is taken.

ii Use an appropriate approximating distribution to find the probability that the total number of plants with the defect in the two samples together is more than 3 and less than 7.   [3]

Cambridge International AS & A Level Mathematics 9709 Paper 71 Q2 Nov 2014

4 Sami claims that he can read minds. He asks each of 50 people to choose one of the 5 letters A, B, C, D or E. He then tells each person which letter he believes they have chosen. He gets 13 correct. Sami says: "This shows that I can read minds, because 13 is more than I would have got right if I were just guessing."

i State null and alternative hypotheses for a test of Sami's claim.   [1]

ii Test at the 10% significance level whether Sami's claim is justified.   [5]

Cambridge International AS & A Level Mathematics 9709 Paper 71 Q2 June 2015

5 At the last election, 70% of people in Apoli supported the president. Luigi believes that the same proportion support the president now. Maria believes that the proportion who support the president now is 35%. In order to test who is right, they agree on a hypothesis test, taking Luigi's belief as the null hypothesis. They will ask 6 people from Apoli, chosen at random, and if more than 3 support the president they will accept Luigi's belief.

i Calculate the probability of a Type I error.   [3]

ii If Maria's belief is true, calculate the probability of a Type II error.   [3]

iii In fact 2 of the 6 people say that they support the president. State which error, Type I or Type II, might be made. Explain your answer. [2]

Cambridge International AS & A Level Mathematics 9709 Paper 71 Q6 Nov 2013

6 The times taken by students to complete a task are normally distributed with standard deviation 2.4 minutes. A lecturer claims that the mean time is 17.0 minutes. The times taken by a random sample of 5 students were 17.8, 22.4, 16.3, 23.1 and 11.4 minutes. Carry out a hypothesis test at the 5% significance level to determine whether the lecturer's claim should be accepted. [5]

Cambridge International AS & A Level Mathematics 9709 Paper 71 Q2 June 2013

7 An examination consists of a written paper and a practical test. The written paper marks (M) have mean 54.8 and standard deviation 16.0. The practical test marks (P) are independent of the written marks and have mean 82.4 and standard deviation 4.8. The final mark is found by adding 75% of M to 25% of P. Find the mean and standard deviation of the final marks for the examination. [3]

Cambridge International AS & A Level Mathematics 9709 Paper 71 Q2 June 2012

8 Heights of a certain species of animal are known to be normally distributed with standard deviation 0.17 m. A conservationist wishes to obtain a 99% confidence interval for the population mean, with total width less than 0.2 m. Find the smallest sample size required. [4]

Cambridge International AS & A Level Mathematics 9709 Paper 71 Q2 November 2013

9 a The time for which Lucy has to wait at a certain traffic light each day is T minutes, where T has probability density function given by:

$$f(t) = \begin{cases} \dfrac{3}{2}t - \dfrac{3}{4}t^2 & 0 \leqslant t \leqslant 2, \\ 0 & \text{otherwise.} \end{cases}$$

Find the probability that, on a randomly chosen day, Lucy has to wait for less than half a minute at the traffic light. [3]

b

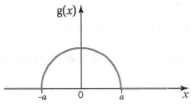

The diagram shows the graph of the probability density function, g, of a random variable X. The graph of g is a semicircle with centre $(0, 0)$ and radius a. Elsewhere $g(x) = 0$.

i Find the value of a. [2]

ii State the value of $E(X)$. [1]

iii Given that $P(X < -c) = 0.2$, find $P(X < c)$. [2]

Cambridge International AS & A Level Mathematics 9709 Paper 71 Q3 Nov 2014

10 The random variable X has a probability density function given by:

$$f(x) = \begin{cases} a - \sec^2 x & 0 \leq x \leq \frac{\pi}{4}, \\ 0 & \text{otherwise.} \end{cases}$$

i Show that $a = \frac{8}{\pi}$.

ii Find $P\left(0 \leq X \leq \frac{\pi}{6}\right)$

11 In a survey a random sample of 150 households in Nantville were asked to fill in a questionnaire about household budgeting.

i The results showed that 33 households owned more than one car. Find an approximate 99% confidence interval for the proportion of all households in Nantville with more than one car. [4]

ii The results also included the weekly expenditure on food, x dollars, of the households. These were summarised as follows.

$$n = 150 \qquad \sum x = 19\,035 \qquad \sum x^2 = 4\,054\,716$$

Find unbiased estimates of the mean and variance of the weekly expenditure on food of all households in Nantville. [3]

iii The government has a list of all the households in Nantville numbered from 1 to 9526. Describe briefly how to use random numbers to select a sample of 150 households from this list. [3]

Cambridge International AS & A Level Mathematics 9709 Paper 71 Q4 Nov 2014

12 Jagdeesh measured the lengths, x minutes, of 60 randomly chosen lectures. His results are summarised below.

$$n = 60 \qquad \sum x = 3420 \qquad \sum x^2 = 195\,200$$

i Calculate unbiased estimates of the population mean and variance. [3]

ii Calculate a 98% confidence interval for the population mean. [3]

Cambridge International AS & A Level Mathematics 9709 Paper 71 Q3 Nov 2015

13 The score on one throw of a 4-sided die is denoted by the random variable X with probability distribution as shown in the table.

x	0	1	2	3
$P(X = x)$	0.25	0.25	0.25	0.25

i Show that $\text{Var}(X) = 1.25$. [1]

The die is thrown 300 times. The score on each throw is noted and the mean, $\bar{X}$, of the 300 scores is found.

ii Use a normal distribution to find $P(\bar{X} < 1.4)$. [3]

iii Justify the use of the normal distribution in part **ii**. [1]

Cambridge International AS & A Level Mathematics 9709 Paper 71 Q5 June 2014

14 The random variable X has a probability density function given by:

$$f(x) = \begin{cases} \dfrac{ax^2}{2x^3 - 1} & 1 \leqslant x \leqslant 2, \\ 0 & \text{otherwise.} \end{cases}$$

  i Show that $a = \dfrac{6}{\ln 15}$.

  ii The 15th percentile of X is denoted by p. Find the value of p correct to 3 significant figures.

15 Weights of cups have a normal distribution with mean 91 g and standard deviation 3.2 g. Weights of saucers have an independent normal distribution with mean 72 g and standard deviation 2.6 g. Cups and saucers are chosen at random to be packed in boxes, with 6 cups and 6 saucers in each box. Given that each empty box weighs 550 g, find the probability that the total weight of a box containing 6 cups and 6 saucers exceeds 1550 g. **[5]**

Cambridge International AS & A Level Mathematics 9709 Paper 71 Q3 June 2013

16 A cereal manufacturer claims that 25% of cereal packets contain a free gift. Lola suspects that the true proportion is less than 25%. In order to test the manufacturer's claim at the 5% significance level, she checks a random sample of 20 packets.

  i Find the critical region for the test. **[5]**

  ii Hence find the probability of a Type I error. **[1]**

  Lola finds that 2 packets in her sample contain a free gift.

  iii State, with a reason, the conclusion she should draw. **[2]**

Cambridge International AS & A Level Mathematics 9709 Paper 71 Q4 Nov 2012

17 Kieran and Andreas are long-jumpers. They model the lengths, in metres, that they jump by the independent random variables $K \sim N(5.64, 0.0576)$ and $A \sim N(4.97, 0.0441)$ respectively. They each make a jump and measure the length. Find the probability that

  i the sum of the lengths of their jumps is less than 11 m **[4]**

  ii Kieran jumps more than 1.2 times as far as Andreas. **[6]**

Cambridge International AS & A Level Mathematics 9709 Paper 71 Q7 Nov 2013

18 A random variable X has the distribution Po(3.2).

  a A random value of X is found.

    i Find $P(X \geqslant 3)$. **[2]**

    ii Find the probability that $X = 3$ given that $X \geqslant 3$. **[3]**

  b Random samples of 120 values of X are taken.

    i Describe fully the distribution of the sample mean. **[2]**

    ii Find the probability that the mean of a random sample of size 120 is less than 3.3. **[3]**

Cambridge International AS & A Level Mathematics 9709 Paper 71 Q5 June 2012

19 A publishing firm has found that errors in the first draft of a new book occur at random and that, on average, there is 1 error in every 3 pages of a first draft. Find the probability that in a particular first draft there are

  i exactly 2 errors in 10 pages   **[2]**

  ii at least 3 errors in 6 pages   **[3]**

  iii fewer than 50 errors in 200 pages.   **[4]**

Cambridge International AS & A Level Mathematics 9709 Paper 71 Q6 June 2015

20 A random variable X has probability density function given by:

$$f(x) = \begin{cases} \dfrac{k}{x} & 1 \leqslant x \leqslant a, \\ 0 & \text{otherwise,} \end{cases}$$

where k and a are positive constants.

  i Show that $k = \dfrac{1}{\ln a}$.   **[3]**

  ii Find E(X) in terms of a.   **[3]**

  iii Find the median of X in terms of a.   **[4]**

Cambridge International AS & A Level Mathematics 9709 Paper 71 Q7 June 2014

21 At a certain hospital it was found that the probability that a patient did not arrive for an appointment was 0.2. The hospital carries out some publicity in the hope that this probability will be reduced. They wish to test whether the publicity has worked.

  i It is suggested that the first 30 appointments on a Monday should be used for the test. Give a reason why this is not an appropriate sample.   **[1]**

A suitable sample of 30 appointments is selected and the number of patients that do not arrive is noted. This figure is used to carry out a test at the 5% significance level.

  ii Explain why the test is one-tail and state suitable null and alternative hypotheses.   **[2]**

  iii State what is meant by a Type I error in this context.   **[1]**

  iv Use the binomial distribution to find the critical region, and find the probability of a Type I error.   **[5]**

  v In fact 3 patients out of the 30 do not arrive. State the conclusion of the test, explaining your answer.   **[2]**

Cambridge International AS & A Level Mathematics 9709 Paper 71 Q7 Nov 2015

22 The weights, in kilograms, of men and women have the distributions $M \sim N(78, 7^2)$ and $W \sim N(66, 5^2)$ respectively.

  i The maximum load that a certain cable car can carry safely is 1200 kg. If 9 randomly chosen men and 7 randomly chosen women enter the cable car, find the probability that the cable car can operate safely.   **[5]**

  ii Find the probability that a randomly chosen woman weighs more than a randomly chosen man.   **[4]**

Cambridge International AS & A Level Mathematics 9709 Paper 71 Q6 Nov 2015

23 The independent variables X and Y are such that $X \sim \text{B}(10, 0.8)$ and $Y \sim \text{Po}(3)$. Find

   i $\text{E}(7X + 5Y - 2)$ **[2]**

   ii $\text{Var}(4X - 3Y + 3)$ **[4]**

   iii $\text{P}(2X - Y = 18)$.   **[4]**

Cambridge International AS & A Level Mathematics 9709 Paper 71 Q7 June 2015

24 The random variable X has a probability density function given by:

$$f(x) = \begin{cases} e^{2x-2} & 1 \leqslant x \leqslant k, \\ 0 & \text{otherwise.} \end{cases}$$

   i Show that $k = \dfrac{2 + \ln 3}{2}$

   ii The median of X is denoted by m. Find the value of m.

25 The lengths, x m, of a random sample of 200 balls of string are found and the results are summarised by $\sum x = 2005$ and $\sum x^2 = 20175$.

   i Calculate unbiased estimates of the population mean and variance of the lengths.   **[3]**

   ii Use the values from part **i** to estimate the probability that the mean length of a random sample of 50 balls of string is less than 10 m.   **[3]**

   iii Explain whether or not it was necessary to use the Central Limit theorem in your calculation in part **ii**.   **[2]**

Cambridge International AS & A Level Mathematics 9709 Paper 71 Q4 June 2013

26 At work Jerry receives emails randomly at a constant average rate of 15 emails per hour.

   i Find the probability that Jerry receives more than 2 emails during a 20-minute period at work.   **[3]**

   ii Jerry's working day is 8 hours long. Find the probability that Jerry receives fewer than 110 emails per day on each of 2 working days.   **[4]**

   iii At work Jerry also receives texts randomly and independently at a constant average rate of 1 text every 10 minutes. Find the probability that the total number of emails and texts that Jerry receives during a 5-minute period at work is more than 2 and less than 6.   **[4]**

Cambridge International AS & A Level Mathematics 9709 Paper 71 Q7 June 2012

27 The volumes of juice in bottles of Apricola are normally distributed. In a random sample of 8 bottles, the volumes of juice, in millilitres, were found to be as follows:

| 332 | 334 | 330 | 328 | 331 | 332 | 329 | 333 |

    i Find unbiased estimates of the population mean and variance.   **[3]**

A random sample of 50 bottles of Apricola gave unbiased estimates of 331 millilitres and 4.20 millilitres^2 for the population mean and variance respectively.

    ii Use this sample of size 50 to calculate a 98% confidence interval for the population mean.   **[3]**

    iii The manufacturer claims that the mean volume of juice in all bottles is 333 millilitres. State, with a reason, whether your answer to part **ii** supports this claim.   **[1]**

Cambridge International AS & A Level Mathematics 9709 Paper 71 Q4 Nov 2011

 28 Customers arrive at an enquiry desk at a constant average rate of 1 every 5 minutes.

    i State one condition for the number of customers arriving in a given period to be modelled by a Poisson distribution.   **[1]**

Assume now that a Poisson distribution is a suitable model.

    ii Find the probability that exactly 5 customers will arrive during a randomly chosen 30-minute period.   **[2]**

    iii Find the probability that fewer than 3 customers will arrive during a randomly chosen 12-minute period.   **[3]**

    iv Find an estimate of the probability that fewer than 30 customers will arrive during a randomly chosen 2-hour period.   **[4]**

Cambridge International AS & A Level Mathematics 9709 Paper 71 Q6 Nov 2011

 29 A random variable X has the distribution Po(1.6).

    i The random variable R is the sum of three independent values of X. Find $P(R<4)$.   **[3]**

    ii The random variable S is the sum of n independent values of X. It is given that

$$P(S=4) = \frac{16}{3} \times P(S=2).$$

      Find n.   **[4]**

    iii The random variable T is the sum of 40 independent values of X. Find $P(T>75)$.   **[4]**

Cambridge International AS & A Level Mathematics 9709 Paper 71 Q7 Nov 2012

 30 The random variable X has a probability density function given by:

$$f(x) = \begin{cases} 2p\cos^2 x - p & 0 \leqslant x \leqslant \frac{\pi}{4}, \\ 0 & \text{otherwise.} \end{cases}$$

    i Show that $2p\cos^2 x - p \equiv p\cos 2x$.

    ii Hence, show that $p=2$.

    iii Find $P\left(0 \leqslant X \leqslant \frac{\pi}{6}\right)$.

Extension Questions

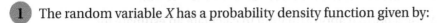

1. The random variable X has a probability density function given by:

$$f(x) = \begin{cases} e^x e^{(k-1)x} & a \leqslant x \leqslant b, \\ 0 & \text{otherwise,} \end{cases}$$

where $k \in \mathbb{N}$ and $b > a$, $a, b \in \mathbb{R}$.

i Show that $e^{kb} - e^{ka} = k$.

ii Hence, or otherwise, determine whether or not each of the following substitutions for a, b and k make $f(x)$ a valid probability density function.

$k = 2$	$a = 0$	$b = \ln\sqrt{3}$
$k = 3$	$a = 0$	$b = \ln 2$
$k = 2$	$a = \ln\sqrt{14}$	$b = \ln 4$
$k = -2$	$a = -\dfrac{33}{32}$	$b = \ln 4$

2. A random variable X has the Poisson distribution $\text{Po}(\lambda)$.

i Given that $P(X=2) = 0.1$, show that $\lambda = \sqrt{0.2 e^\lambda}$.

ii Hence use the iterative formula $\lambda_{n+1} = \sqrt{0.2 e^{\lambda_n}}$ with $\lambda_0 = 4$ to find a possible value of the parameter correct to 3 significant figures.

iii Hence find an estimate for $P(X<3)$.

3. X, Y and Z are independent random variables, where $X \sim \text{B}(n, 0.5)$, $Y \sim \text{Po}(n)$ and $Z \sim \text{N}(n, n)$. The random variable W is a linear combination of X, Y and Z, where $W = 2X + 3Y + 4Z$.

i Given that $E(W)^2 - 2\text{Var}(W) = 420$, find the value of n.

ii Find the probability that $X = Y = 1$.

4. The random variables X_1 and X_2 have probability density functions given by

$$f(x) = \begin{cases} 2x - a & 0 \leqslant x \leqslant b, \\ 0 & \text{otherwise.} \end{cases}$$

$$g(x) = \begin{cases} 12x - b & 0 \leqslant x \leqslant a, \\ 0 & \text{otherwise.} \end{cases}$$

Find the value of a and the value of b.

5. When people fly on commercial airlines, their luggage is usually weighed, but the passengers are not weighed.

i Explain why the weight of passengers does not need to be measured.

ii For larger planes with a greater number of passengers, does it become more or less important to consider the weight of individual passengers?

iii If the New Zealand rugby team (25 players) boards a small plane that can carry 40 passengers, is the Central Limit Theorem still valid?

6 The random variable X has a Poisson distribution, where $X \sim \text{Po}(\lambda)$.

  i In the case where $\lambda = 4$, use the Poisson formula to calculate $P(2 \leqslant X \leqslant 5)$.

  ii In the case where $\lambda = 4$, use a normal approximation to calculate an estimate for $P(2 \leqslant X \leqslant 5)$.

  iii In the case where $\lambda = 4$, find magnitude of the percentage error generated by using a normal approximation to calculate an estimate for $P(2 \leqslant X \leqslant 5)$.

  iv As the value of λ increases, describe what you would expect to happen to the magnitude of the percentage error generated by using a normal approximation for a Poisson distribution to find $P(2 \leqslant X \leqslant 5)$.

7 The number of cars crossing a bridge in a 10-minute interval can be modelled by a Poisson distribution with a mean of λ. The probability that the first car crosses the bridge between 10 minutes and 20 minutes is p.

  i Show that $e^{\lambda} - pe^{2\lambda} = 1$.

  ii By forming a suitable quadratic equation, or otherwise, find the range of possible values of p for which the model is valid.

8 In 2008, the masses of a large number of stones were known to be normally distributed with mean 10 kg and standard deviation 0.5 kg. In 2018, a scientist suspects that the stones have eroded over time and so their masses have reduced. The scientist takes a sample of n stones and discovers that their mean mass is μ kg.

  i In the case where $n = 16$, find the maximum value of μ if there is evidence to support the scientist's suspicion at the 5% significance level.

  ii In the case where the value of n is unknown, find an expression for the revised maximum value of μ if there is evidence to support the scientist's suspicion at the 1% significance level.

9 The random variable X has a Poisson distribution, where $X \sim \text{Po}(x)$. The random variable Y has a binomial distribution, where $Y \sim \text{B}(2, e^{-x})$ and $x > 0$.

  Find the exact value of x such that $P(X = 0) = P(Y = 1)$.

10 Antonio has a standard 6-sided die that he believes is biased. He rolls the die 5 times and he rolls a '5' twice.

  i Carry out a hypothesis test at the 10% level to show there is insufficient evidence that Antonio's die is biased.

  After 10 rolls of the die, Antonio has rolled a '5' on four occasions.

  ii Carry out a new hypothesis test at the 10% level to show there now is sufficient evidence that Antonio's die is biased.

  iii Explain why Antonio rolling a 5 on 40% of the rolls can be insignificant after 5 rolls, but significant after 10 rolls.

11 The weights of sacks of potatoes from a particular factory are normally distributed with mean μ and standard deviation 0.5 kg. A random sample of n sacks is taken and a 99% confidence interval for μ is calculated. A second random sample of $(n-10)$ sacks is taken and a 95% confidence interval for μ is calculated. The 99% confidence interval is found to be m times larger than the 95% confidence interval.

  i Given that $m = 1.30001509$ (to 8 d.p.), calculate the value of n.

  ii For different sized random samples with equal means, find the conditions under which it is possible for the magnitude of a 95% confidence interval to be larger than the magnitude of a 99% confidence interval.

12 The number of children visiting a fairground each hour can be modelled by the random variable X, where $X \sim \text{Po}(5)$.

The probability that a child visiting the fairground uses the rollercoaster ride is $\frac{4}{5}$ and the decision of each child to use the rollercoaster is independent. In the case where $2n$ children visit the fairground, find an expression in terms of n for the probability that exactly half of them use the rollercoaster. Write your expression as a fraction in its simplest terms.

GLOSSARY

acceptance region The values of X which lead to H_0 being accepted. These are all the values of X not in the critical region.

alternative hypothesis The alternative hypothesis, H_1, states what you are attempting to prove.

bias Anything that occurs in the process of sampling which prevents the sample from being representative of the whole population.

binomial distribution A discrete probability distribution in which there are only two outcomes, success and failure, to any trial, the probability, p, of success is the same in each trial and there are a fixed number of n independent trials.

central limit theorem The central limit theorem states that, if the sample size is large enough, the sampling distribution of the sample mean is normally distributed, even if the population is not normally distributed.

combination A combination is an arrangement of objects chosen from a given set where order does not matter. The number of arrangements of r different objects chosen from a set of n objects where order does not matter is written as ^nC_r or $\binom{n}{r}$, the formula for ^nC_r is

$$\frac{n!}{r!(n-r)!}$$

confidence interval A range of values which is likely to contain a population parameter and is calculated from sample data. For example, a confidence interval for a population mean will be based on the sample mean ± an error term.

confidence level The probability that the population parameter falls within the confidence interval. It is given as a percentage.

continuity correction When a discrete probability distribution is approximated by a continuous probability distribution a continuity correction is needed. For a discrete random variable $P(X = x)$ may have a value other than 0, but for a continuous random variable $P(X = x) = 0$ for all individual values.

continuous random variable A random variable that can take infinitely many values.

critical value The critical values are the boundaries of the rejection (critical) region. In the upper tail of a distribution, the smallest value of X to fall into the rejection region is known as the critical value. In the lower tail of a distribution, the critical value is the largest value of X to fall into the rejection region.

expectation The expectation of a random variable X is what you would expect to get if you took a large number of values of X and found their mean. The mean value is called the expected value and is written $E(X)$ or μ. $E(X) = \Sigma xp(x)$.

linear Linear means of a straight line. $y = mx + c$ is the equation of a straight line so any expression of the form $mx + c$ is said to be linear.

mean The mean $(\bar{x})$ is an example of an average. It is the sum of all the values divided by the total number of values. For a simple set of n numbers the mean is $\bar{x} = \dfrac{\Sigma x}{n}$, and for a frequency distribution the mean is $\bar{x} = \dfrac{\Sigma fx}{\Sigma f}$.

normal distribution Normal distributions occur in nature. They are the probability distribution for a continuous random variable, for example, the length of leaves for a particular type of tree. Values which are closer to the mean value have a higher probability than values further away from the mean value. When represented graphically, normal distributions have a bell shaped curve and are symmetrical about the mean.

null hypothesis A null hypothesis, H_0, states the expected or theoretical outcome.

one-tailed test In a one-tailed test you use one 'tail' end of the probability distribution. The alternative hypothesis specifies whether the parameter you are investigating is either greater than or less than the value you are using in H_0 as you are looking for a definite increase or decrease in the parameter.

piecewise function A probability density function (p.d.f.) for a continuous random variable is usually expressed as a set of functions, each of which is defined for a particular range of values of the random variable. This set of functions is called a piecewise function.

point estimate The statistic obtained from the sample that is used as a simple estimate of the population parameter. For example, the sample mean can be used as a point estimate of the population mean.

poisson distribution When a discrete random variable is used to model situations where the number of times an event occurs in a given interval. For example, the number of phone calls received in a 20 minute interval. It is used when events occur one at a time, independently of each other, randomly and at a constant mean rate.

population parameter   A number that summarises information about a population, such as the mean of the population.

population proportion   The fraction of a population that possess a particular attribute; for example, the fraction of people in a population who own a car.

probability   The measure of the likelihood that the outcome or event will occur as a result of an experiment.

probability density function   The probability density function (p.d.f.) of the continuous random variable X is the function $f(x)$ such that $P(a \leqslant X \leqslant b) = \int_a^b f(x)\,dx$.

rejection region   As an alternative to calculating the p-value, it can be useful to calculate the values of X that would lead to H_0 being rejected. These values are known as the rejection region (or the critical region).

sampling distribution   The probability distribution of a statistic. For example, the sampling distribution of the sample mean is the theoretical set of all possible values the sample mean can take.

significance level   The probability of rejecting H_0 when it is true.

standard deviation   Standard deviation, σ, is used as a measure of variation. It is the square root of the variance. $\sigma = \sqrt{\dfrac{\sum(x - \bar{x})^2}{n}}$ and $\sigma = \sqrt{\dfrac{\sum x^2}{n} - \bar{x}^2}$ are equivalent formulae for finding the standard deviation.

standard error   The standard deviation of the sampling distribution of a statistic is called the standard error of the statistic.

statistic   A number that summarises information about a population, calculated from a sample such as the sample mean.

test statistic   The statistic that is used for the hypothesis test.

two-tailed test   In a two-tailed test you use both 'tail' ends of the distribution. The alternative hypothesis states the parameter is not equal to the value in the null hypothesis as you are just looking for a change in the parameter.

Type I error   A Type I error is made when H_0 is rejected when it is true. The probability of a Type I error is the same as the significance level of the test.

Type II error   A Type II error is made when H_0 is accepted when it is false. You can also think of this as H_0 being accepted when H_1 is true.

variance   Variance, σ^2, is used as a measure of variation. The deviation (difference) of each value from the mean is found, squared, and then the mean of these squared deviations is found. $\sigma^2 = \dfrac{\sum(x - \bar{x})^2}{n}$ and $\sigma^2 = \dfrac{\sum x^2}{n} - \bar{x}^2$ are equivalent formulae for finding the variance. For frequency distributions the formulae are $\sigma^2 = \dfrac{\sum f(x - \bar{x})^2}{\sum f}$ and $\sigma^2 = \dfrac{\sum fx^2}{\sum f} - \bar{x}^2$. For probability distributions the formulae are

$\sigma^2 = \text{Var}(X) = \sum x^2 p - (E(X))^2$ or $\sigma^2 = \text{Var}(X) = E(X^2) - (E(X))^2$.

INDEX